Workbench Guide to Tape Recorder Servicing

Workbench Guide to Tape Recorder Servicing

G. Howard Poteet

Parker Publishing Co.
West Nyack, New York

PARKER PUBLISHING COMPANY, INC.
West Nyack, New York

Library of Congress Cataloging in Publication Data

Poteet, G Howard.
Workbench guide to tape recorder servicing.

Bibliography: p.
Includes index.
1. Magnetic recorders and recording--Repairing.
I. Title.
TK7881.6.P6 621.389'32 76-22201
ISBN 0-13-965293-0

Printed in the United States of America

A Word From the Author On the Practical Value This Book Offers

Today, each home boasts a tape recorder—sometimes several. However, many of them function badly, distorting what should be sparkling stereo sound, chewing up tape containing the baby's first words, or failing to erase the sounds of previous recordings. Tape equipment needs servicing and service requires know-how.

Now you can become an expert in an area few people have mastered because this book tells you in plain and simple terms how you can most effectively use, service and repair any tape recorder. Further, it tells you how to record, edit and repair any type of tape on the market—open-reel, cassette or cartridge. It gives you ways to do it that will make you a tape wizard quickly, and without wading through an assortment of technical manuals.

Specifically, the book tells you how to service *wow* and *flutter* out of your tape equipment so you can enjoy Bach or bluegrass. Not only will you be able to select the best equipment for your needs, but you'll learn new ways to use equipment you already own.

This book will be invaluable for the technician or hobbyist who needs to know precise methods of working with complex equipment. Even the beginning experimenter will have no trouble in following each of the specific guidelines because of the nontechnical chapters. You will find all the suggestions are based on the best available research which has been translated into reasonable and sound applications that really work. We have even included the actual repair manuals issued by representative manufacturers.

Want to find out why that cassette tape won't move—in minutes? Simply turn to Chapter 5. Want to know the easy steps in replacing a worn recording head and aligning it according to specifications? You'll find the information in Chapter 7. In addition, you'll discover new ways to track down problems in recording or playback components in any tape equipment. Don't let that recorder gather dust on your workbench—look in this book for information on mechanical enigmas like tape transport mechanisms or electronic puzzlers like bias settings, as well as how to make truly professional sounding recordings.

In your own professional work, you'll be pleased to learn tricks and techniques ranging from how to select the right tape for the job, to being able to know precisely when a recording/playback head needs to be replaced, to building a cheap but good recording studio in a closet.

There are plenty of ways in which you will use the clever ideas in this book. You can, for example, tuneup any recorder and improve its sound by following the suggestions in Chapter 6. There you will find even more amazing ideas which will have your friends coming to you for your expert opinion on how to solve their tape recording problems.

Best of all, you don't have to be a mathematical whiz or an engineering genius to put these ideas into effect. You don't need to waste your valuable time plowing through dull, dry research reports or in costly and time-wasting experimentation with slide rule in hand. This book presents the information you need in simple and concise terms and cuts away the fat of useless theory. All you need do is to follow the simple step-by-step rules and watch your expertise increase.

Find out, for instance, the things schematics don't tell you about tape recorders. Learn where you can get the materials you need to do the job faster and cheaper. Know how to plan for the unexpected.

Are you interested in keeping ahead of the pack by becoming a tape recording wizard? This book will supply you with enough amazing and ingenious schemes to do just that.

Contents

Workbench Guide to Tape Recorder Servicing

Chapter 1

Developing a Practical Understanding of Magnetic Recording

As an electronics technician, you will find the following history of magnetic recording interesting, the simplified theory of magnetism vital, and the suggested uses for tape recordings intriguing. To develop the fullest practical understanding of magnetic recording, you must know something of its origin. You'll find plenty of opportunities to pass this information along (in bits and pieces) to your customers when they ply you with questions. This background, together with the practical technical know-how you will gain from later chapters, will assure them of your comprehensive knowledge of magnetic recording.

Origin of magnetic recording

Ever wonder how magnetic recording originated? Back in 1898, Valdemar Poulsen, a Danish inventor, recorded telephone messages on a fast-moving piano wire to keep transmissions of conversations secret. His Telegraphone (which played for sixty seconds) was the first practical tape recorder.

In following years, the Blattnerphone, the Marconi-Stille Recorder and others used thin steel bands which whizzed through the machines at a terrifying speed of 270 feet a minute! Using one was like recording on a band saw—heaven help you when it broke! A further problem was that splices required a welder.

Although the Magnetophone used tape (invented by Fritz Pfleumer in 1927), most research went into the wire recorder, which reached its peak with inventor Marvin Camras' 1943 device. During World War II, some experimentation with tape was done in the United States; in Germany tape recording technicians developed some very sophisticated equipment.

Eventually, tape (introduced by 3M in the United States in 1947) won out over wire because wire developed kinks, was difficult to splice, and produced poor fidelity. Earliest tapes (which were open-reel) ran through machines at a breathtaking three feet a second—about twelve times as fast as they do today.

The first cassette tape (the Fidelipac, invented by Charles Eash around 1955) and the first cartridge (invented by Bernard Cousino at about the same time) were intended to permit tape recorders to be used in cars, without complicated threading. Cassettes and cartridges were so easy to use that they were adapted for home and business use as well. Initially, each manufacturer used a different size and system; none were compatible. Later, agreements were made to standardize them.

In ensuing years, fantastic progress has been made in creating more sophisticated equipment (Dolby noise reduction units, switchable bias equalizers, for example) and in producing more responsive tape (extended range tapes, CrO_2 tapes). However, the process of magnetic recording is still basically the same system developed by Poulsen in 1898.

Theory of magnetism simplified

Magnetic recording is based on simple magnetic theory. Each magnet has a north pole and a south pole. If two bar magnets are placed end to end, the two north poles (or the two south poles) will repel each other. If a north pole and a south pole are placed together, they will attract each other. Thus we have the law: Like repels, unlike attracts.

The area between these two forces at the end of the poles of the magnets is called the magnetic field. Metals placed in this magnetic field will become magnetized; that is, their individual molecules will become magnets. "Hard" magnetic metals will remain magnetized after being removed from the field; "soft" magnetic metals will not.

Although metals (ranging from "soft" metals like Mu-metal or permalloy to "hard" metals like iron, Alnico, or steel) can be mag-

Figure 1-1. Recorded sound track made visible on a magnetic audible range tape by magnification of iron filings sprinkled on top of the iron oxide surface. (Courtesy of 3M)

netized by holding them close to a magnet, magnetism can also be produced electrically by wrapping an insulated electric wire around a soft metal bar (which thus will not be permanently magnetized). As long as current flows through the wire, the bar picks up magnetism by induction. The magnet's power may be increased by one or more of the following three ways: by increasing the current flowing through the wire, by increasing the number of coils of wire wrapped around the metal bar, by using different types of metal as the core.

Understanding that the magnetic force or flux created by electrical force flowing through a coil can be varied, and that we can magnetize other metals ("hard" metals retain magnetization; "soft" metals do not), we can put these theories to work.

Practical application of magnetism in recorders

Sound waves are converted into electrical energy by a microphone. These electrical signals are converted into magnetism by sending them through a coil of wire wrapped around a "soft" iron bar (which does not retain magnetization). A tape (a strip of plastic coated with "hard" iron particles which retain magnetization is passed through the magnetic field created by the iron bar and is magnetized.

If a piece of magnetic tape is moved past the recording head or iron bar at 3-3/4 ips (inches per second) and the microphone is sending electrical impulses at the rate of 1000 per second (referred to as 1000 cps), one thousand magnets are formed in one second on that 3-3/4 inches of tape. The intensity of the sound wave hitting the microphone would result in varying numbers of magnets being formed on the tape (500 signals per second would result in 500 magnets being formed in that 3-3/4 inches of tape).

Playback theory is just as simple. Electrical current is created by moving a magnet through a coil. Tape which contains many magnets (as we have just seen) is moved through a coil (the playback head) creating a series of electrical impulses. These electrical impulses are then amplified through an amplifier and a loudspeaker.

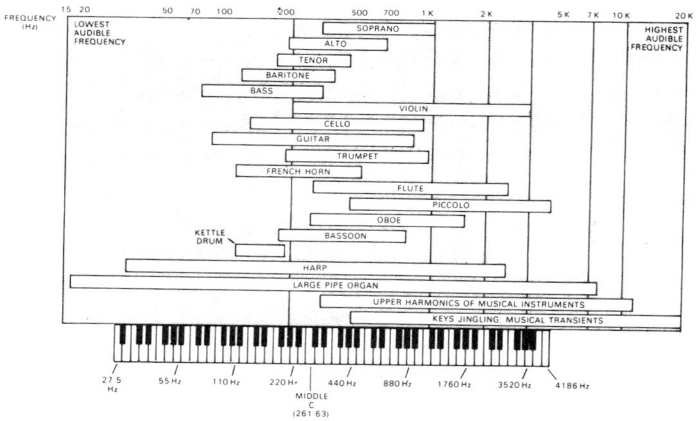

Figure 1-2. Frequency ranges of various sounds. (Courtesy of 3M)

Standard procedures for recording

Recording (putting magnetic impulses onto tape) is done in much the same way with any recorder. Turn the equipment on, place in record mode, and speak into the microphone (or connect to a radio or record player with appropriate cables). Most recorders have automatic controls which are not affected by external bass or treble knobs, but to play it safe, turn them flat (in the middle position between high and low) to avoid distortion.

Although many recorders have automatic level controls, it is often necessary to ride gain (adjust the volume manually) while watching the meter. Some cheaper recorders do not have meters; others have devices similar to a car's idiot lights. Many recorders permit you to listen as you record and thus you can adjust the volume by ear.

Tell your customers to remember that recording is as easy as 1-2-3. One, turn the recorder on; two, put it into record mode; three, watch the volume.

As a serviceman, you are well aware that many problems your customers present you with are the result of their not understanding how to operate their equipment. Although it seems obvious, customers have often been unable to record simply because they have put their recorder into playback mode, they have the volume down too low, or they have forgotten to connect the microphone.

Figure 1-3. A simple tape recorder. (Courtesy of General Electric)

Figure 1-4. A highly complex recorder. (Courtesy of Hervic)

Figure 1-5. A typical tape recorder/player circuit.

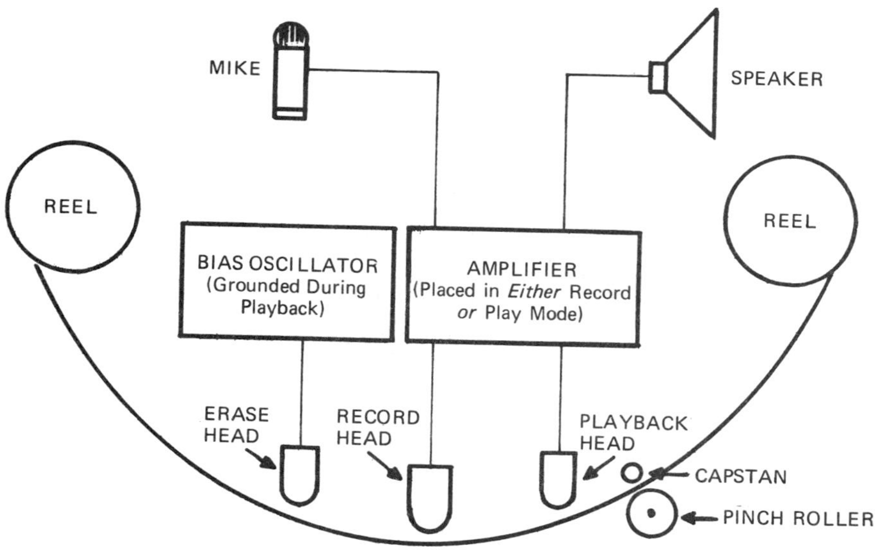

Methods of playback: a summary

Tape is played back (after rewinding) simply by pressing the playback mode switch.

The playback head receives the electromagnetic responses from the tape as it moves past, turns these magnetic signals into electric signals which are fed into an amplifier which increases their strength and feeds them into a loud-speaker so they can be heard at room level. The listener needs only to adjust the volume level to suit his needs.

Remind your customers to think of 1-2-3 again. One, turn the player on; two, put it into the playback mode (customers often mistakenly press the record switch instead, thus erasing the tape); three, adjust the volume.

Ways to use tape

Suggest some of the many uses for tape recordings to your

Figure 1-6. An open-reel, a cassette, and a cartridge tape. (Courtesy of 3M)

Figure 1-7. Tape recording offers both pleasure and profits. (Courtesy of 3M)

customers when you are getting them to trade in old equipment for new.

Tape recordings can be used in business (recording phone calls, delivering canned sales pitches, keeping areas under surveillance), pleasure (playing pre-recorded tapes, recording music or dramas off the air or live, taping big moments in the family's life), or self-improvement (improving your diction in speeches, listening and learning from taped books and lectures, recording and criticizing your sales methods). Uses for tape recorders and tape recordings are practically unlimited.

Chapter 2

Evaluating Tape Equipment

As a serviceman, you must know the types of equipment available, the ways in which they can be used, and how to judge their quality.

Understanding the need for different equipment

Modern recording and playback equipment use either open-reel, cassette, or cartridge tapes.

Open-reel recorders (also called reel-to-reel) use reels of 50 to 2400 feet of 1/4 inch tape. Professional work is usually done on open-reel equipment because more control over the results is possible—speeds can be varied from 15/16 ips to 15 ips on many machines (for longer play or better fidelity) and the tape is easy to edit. However, open-reel equipment is usually heavier and bulkier than cassette or cartridge equipment.

Cassettes (approximately 4 inches by 2-1/2 inches by 3/8 inch) hold 150 to 600 feet of 1/8 inch tape which travels at only one speed, 1-7/8 ips. Cassette tapes provide both monophonic and stereophonic sound rivaling open-reel recordings, without the inconvenience of threading tape through the machine. But it is almost impossible to edit the tape mechanically (by cutting it and splicing it); it must be done electronically. Also, cassettes are easily damaged and subject to jamming. Tape quality can be impaired by something as small as your thumbprint.

Cartridge players (used in cars and in some home entertainment systems) play 4 track and 8 track stereo exclusively. Cartridge tapes (1/4 inch wide) run through the machine in one direction only at 3-3/4 ips. Like cassettes, they are almost impossible to edit

Figure 2-1. An open-reel recorder (Courtesy of Radio Shack, a Tandy Corporation Co.)

mechanically. In addition, it is impossible to rewind a tape or to put it in the fast forward mode—tapes must be played to the end and then started over again if you wish to rehear a selection. Few cartridge recorders are on the market. Unlike open-reel and cassette recorder/player combinations, most cartridge machines are players only.

Each of these three systems is perfect for some purposes and unacceptable for others. Eight track cartridge systems are ideal for automobiles; they do not require threading, are easily handled, and play continuously. Cassettes are excellent for portable use (cassette equipment is compact), superb for dictation (they do not become unwound easily), and fine for personal listening (they are the simplest to use). Open-reel tapes are best for permanent or semipermanent installations when high fidelity is critical and space is not.

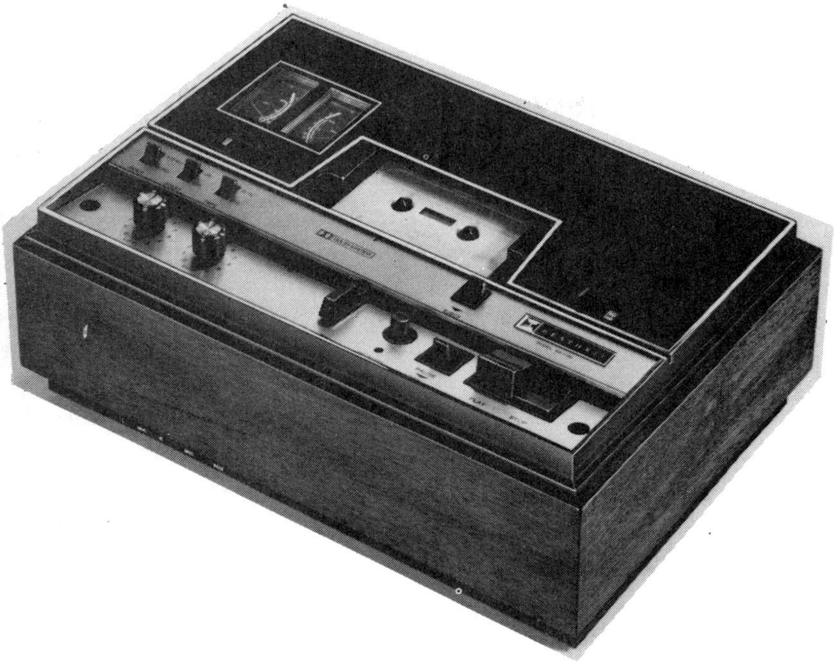

Figure 2-2. A cassette recorder/player. (Courtesy of Heathkit)

Figure 2-3. A cartridge player. (Courtesy of Radio Shack, a Tandy Corporation Co.)

How to find the best equipment for your customer's needs

Since there are hundreds of machines on the market, customers want you to help them make up their minds before they make a purchase. Do that by asking a few simple questions. One, how do they intend to use the recorder (in the car, for dictation, for classical music tapes)? Two, how good are they at using electronic equipment (do they enjoy twiddling dials or do they prefer things to be automatic)? Three, do they prefer standard or superior equipment (translated, this means how much do they want to spend)?

Don't talk price; talk quality. People usually spend more than they originally estimate, especially if it is for something that makes life more enjoyable. Top quality tape equipment will offer many hours of listening pleasure; inferior equipment will be tiring to listen to, used less often and thus in the long run (on a dollar per hour-of-use basis) will be more expensive. Encourage your customers to get the best equipment they can; you'll benefit from it.

Help them make a selection by giving them a choice between two items. Need a recorder for the car? That's simple—cassettes or cartridges? If they want to use pre-recorded tapes, tell them to get a cartridge player. (With either type, you can sell them equipment for home listening, too.) Can't decide between two similar models? Suggest they select the one that is more automatic.

The following chart lists some comparisons among types of tape recorders/players currently available.

Comparisons Among Tape Recorder/Players

Description for Comparison	Open-reel	Cassette	Cartridge
Player available?	yes	yes	yes
Recorder available?	yes	yes	seldom
Operating speed(s)	15/16-15 ips	1-7/8 ips	3-3/4 ips
Total playing time	1-720 min.	15-120 min.	45-90 min.
Self-threading?	seldom	yes	yes
Type of tape carrier	3″-7″ reels	cassette	cartridge (loop)

Description for Comparison	Open-reel	Cassette	Cartridge
Tape width	1/4″	1/8″	1/4″
Editing?	yes	yes	difficult
Sound quality	excellent	fine	fine
Ease of operation	often complex	simple	very simple
Pre-recorded tapes available?	few	many	many
Power	batt/ac	batt/ac	batt(auto)/ ac
Deck available?	yes	yes	yes
Size	usually large	compact	compact
Rewind	fast-very fast	fast	no
Professional use?	yes	seldom	no
Durability	excellent	fine to excellent	fine
Price	generally expensive	inexpensive to moderately expensive	inexpensive

Ways to rate a recorder/player—from motor speeds to sound quality

How do you, the serviceman, judge equipment? Amateurs can go on looks; you have to know what's inside. Obviously, you could look up ratings of equipment in *Consumer Reports* or *Tape Deck Quarterly*, but you'll be happier relying on your own knowledge and judgment.

Some parts of your appraisal will be quick and self-evident. Is the equipment well finished or carelessly thrown together? Are the controls clearly marked, well located and sturdy?

Check the motor speeds. Cassette and cartridge machines will not offer a selection of speeds, but better quality open-reel recorder/players will have a range as wide as from 15/16 ips to 15 ips. Most, however, will offer three speeds (usually 1-7/8, 3-3/4, and 7-1/2 ips).

You'll find differences in rewind time ranging from 150 seconds for a 1200-foot tape (medium priced machine) to 60 seconds for a 1200-foot tape (expensive machine). A good rule of thumb is that a fast rewind indicates a good quality machine. It is something few people think about in buying a machine, and an item on which many manufacturers thus cut corners.

Other signs of a quality machine are *three heads* (one of them permits you to listen to the recorded signal a second after it has been recorded, insuring good recording), *three motors* (producing steady tape speeds), *Dolby noise reduction units* or *switches for chromium dioxide tape* (assuring better sound quality), *permalloy heads* (claimed to outlast ferrite heads), *record equalization switches* (to make better copies of disc recordings), and similar features.

You can choose between items like slide-lever volume controls or the conventional knobs as a matter of personal preference. Some essentials, however, are VU meters, a tape counter with a reset, a pause position for editing and cuing, and an automatic end-of-tape shut off. Use the following checklist.

Checklist for Quality
Tape Recorder/Players

Description	Yes	No
Equipment well finished?		
Controls well marked, easy to operate, sturdy?		
Adequate range of motor speeds? (Where applicable)		
Fast rewind?		
Three heads?		
Three motors?		
Dolby system?		
Switch for low noise-high output tapes?		
Permalloy heads or Ferrite heads?		
Record equalization switches?		

(continued) Description	Yes	No
VU meter?		
Tape counter?		
Pause control?		
Automatic shut-off?		

(More than four checks under the *No* column indicates a machine of less than superior quality. More than eight checks under the *No* column indiates a machine of less than standard quality.)

Play a tape to check the range, balance, and general sound quality. Although test tapes are available, try listening to one of your favorite tapes. You'll find that your familiarity with it gives you a good basis for making a judgment.

In summation, select a recorder with a pleasing appearance, durable parts and features to fit your customer's needs, manufactured by a reputable manufacturer (repairing orphans is next to impossible).

How to make sense of manufacturers' specification sheets

Ever wonder what those specifications listed in the catalog mean? Those listings are accurate indications of recorder quality and are simple for you, the electronic technician, to decipher and explain to your customer.

First, look at the *frequency response*. Frequency is measured in Hz (a wave frequency of one cycle per second). A range between 40 Hz and 13 K Hz would thus be an acceptable minimum for an inexpensive recorder. It would reproduce sounds of a symphony orchestra with reasonable clarity. A fine quality machine capable of reproducing lifelike recordings with accurate fidelity should have a minimum range from 30 to 16,000 Hz. Acceptable equipment would have a range between that of the fine quality machine and that of the inexpensive machine. Extremely fine equipment might have a range from 20 to 20,000 Hz.

Second, check the *signal-to-noise ratio*. This is the voltage ratio between the loudest undistorted tone recorded (and reproduced) by a recorder, and the noise reproduced when the signal (not the bias) is zero. This is rated in decibels (dB), the smallest change the human ear can detect in volume. For a fine quality machine, the ratio should be about 50 dB.

Third, check the *distortion factor*. Distortion is any difference between the original sound and that reproduced by a recorder. In this measurement, "distortion" refers to distortion at full level recording. The maximum in a fine quality machine should be 5 percent.

Fourth, look at the *cross-talk* rating. This means the unintentional mixing of the sounds from one channel with the sounds of another. Measured at 1,000 Hz, for a standard quality machine, the measurement should be no more than 60 db. A high quality stereo recorder/player should measure no more than 25 dB.

Fifth, check the speed variation, if listed, for *wow and flutter*. "Wow" is a form of distortion caused by periodic variation in the speed of the tape which causes variation in pitch. "Flutter" is very short rapid variations in tape speed causing similar variations in sound pitch and volume. If speed variation is not minimized, tapes recorded on one machine cannot be played on another without distortion. Further, pre-recorded and commercially produced tapes cannot be played satisfactorily. The accepted tolerance is ± 2 per cent for standard quality recorder/players and ± 1 percent for quality machines.

The following chart lists typical specifications and their interpretations.

How to Understand Manufacturers' Specification Sheets

FINE — DD-2178 SPECIFICATIONS: *Frequency Response:* Regular (iron oxide tape) ± 3dB from 30 Hz to 16 kHz typical. CrO_2 (chromium dioxide) tape; ± 3dB from 20 Hz to 16 kHz typical.

FINE — *Distortion:* Tape dependent; electronics less than 0.2%.

GOOD — *Cross-talk:* Greater than 25 dB at 1 kHz. *Wow and Flutter:* 0.08% at 7-1/2 ips, 0.10% at 3-3/4 ips.

FINE — *Signal-to-noise ratio:* 58 dB. Inputs: Microphone. *Fast Forward/Rewind Time:* Approximately 45 seconds for 1200′ reel. *Tape Speed:* 1-7/8 ips, 3-3/4 ips, 7-1/2 ips.

FAIR — *Solid State Devices:* 57 transistors and 3 JFTE's. *Dimensions:* 10-1/2″H × 9″D × 14″W. *Power Requirements:* 120 volts, 60 Hz, 15 W.

FINE

You'll find that your customers will have to make some compromises and, in fact, will not need the very top quality responses for most purposes. They thus will make compromises with price, size and quality. You'll find it best to stick to three basic models for recommendations and sales—a cheap model for the kids (not much more than a toy), a standard quality machine for family use, and a top quality recorder/player for the discriminating hi-fi stereo fans.

If you know these basic models thoroughly, you'll be able to sell and service them with a minimum of time loss and difficulty. If you keep juggling too many models, you'll never be able to keep them all straight.

Figure 2-7. In this illustration the Sennheiser RV 55 Voltmeter, FO 55 Weighting Network and Filter Unit, and KB Harmonic Distortion Bridge are being used to check the quality of a tape recorder. (Courtesy of Sennheiser Electronic)

A simple foolproof quality test for any tape recorder

For instant results, the proof is in the listening. Even if you are not a musician, you'll hear the quality in a recorder if you use a tape of a piano solo (preferably a classical work such as *Claire de Lune*). Any distortion shows up at once, especially in sustained notes which tend to quaver in machines lacking quality.

Many hi-fi stereo showrooms use tapes of big showy orchestral works with thumping bass notes and light and quick transient noises like the clashing of cymbals. These sounds can be reproduced by relatively cheap equipment. Truly fine sound reproduction depends as much on silence as on deep brass notes. In fact, a poor piece of equipment can give you "listener fatigue." You'll get a headache from listening to the distortion after a short period of time.

Chapter 3

Making Successful Tape Recorder Installations

Carefully chosen types of installations can bring in many extra dollars. Occasionally, you will install a tape recorder in your customer's home, but most of your installations will be in automobiles. In addition, business installations offer opportunities for extra profits.

Judging installation needs

To build profits with today's wide range of electronic gear, sell your customer modular equipment that doesn't need to be built in. He can install it himself, and you need only suggest ways to hide the connecting cables. If the customer insists on making equipment part of a piece of furniture, sell him a self-contained console.

If a home installation requires building in equipment, you can make it profitable only if you keep the procedure simple, as will be subsequently explained. You must charge at least your minimum service call fee (include it in your total price for the equipment). Avoid such installations if you want to specialize in the electronics end of the business, because you will need to draw more on mechanical and carpentry talents than on your skill in electronics. In such cases, arrange for someone to make installations on a piecework basis. You'll come out dollars ahead!

Auto installations are not complex if you use standard components. It is a field for specialists; most general servicemen do it only as a convenience for old customers.

Because the crime rate is soaring, businesses will often pay you to install recorders for surveillance purposes. This is also a highly specialized job (requiring techniques you may not enjoy), but tie-ins with detective bureaus, guard services and security firms can net you extra income.

In all these opportunities, you must choose to perform only the kinds of jobs and the techniques which will make them profitable.

Simplifying installation

Most recorders are designed for standard hookup to any modern stereo receiver or amplifier. Standard pin plugs permit easy hookups; input and output jacks and circuits are identical for both channels. In stereo recording and playback, connections are made to the appropriate jacks of both channels (left channel recorder input to left channel amplifier output, right channel input to right channel output). Although these connections will record monaural as well, connections need to be made to only one of the recorder's input channels; placing the mode switch in "mono" position records the mono signal on both tracks. For mono playback, connections are normally made from the output jacks of both channels to both channel inputs on the amplifier, allowing the mono signal to play through both channels of the stereo system. Such hookup information is always included in the instruction booklet which accompanies recorder/playback equipment. When installation is no more complicated than this, your customers can do it—you would be wasting your valuable time.

When home installation is more complicated, consisting of putting equipment in place, connecting, and hiding wires, it should require no more than one hour of your time. Even then, it may not be profitable. This is also true of business installations. The key to profit is standardization and simplification.

Home and business installation will require concealment of connecting cables. This can be done in three ways: by boring holes and running wire behind walls (often impractical because of the high expense and servicing difficulties), tacking wire or wrapping it in strands or enclosing it in devices, and using the exposed wire in a decorative manner. All require a minimum of tools.

To make successful installations with cables concealed behind the walls, you will need a drill or bit brace, a 5/8 inch bit (or drill), a hammer, a chisel, a fishtape and reel, and a screwdriver.

You will probably only need to conceal the cables connecting the amplifier to the speaker. For most installations, the speakers need to be about eight feet apart. You use the same techniques to install extension speakers in other rooms.

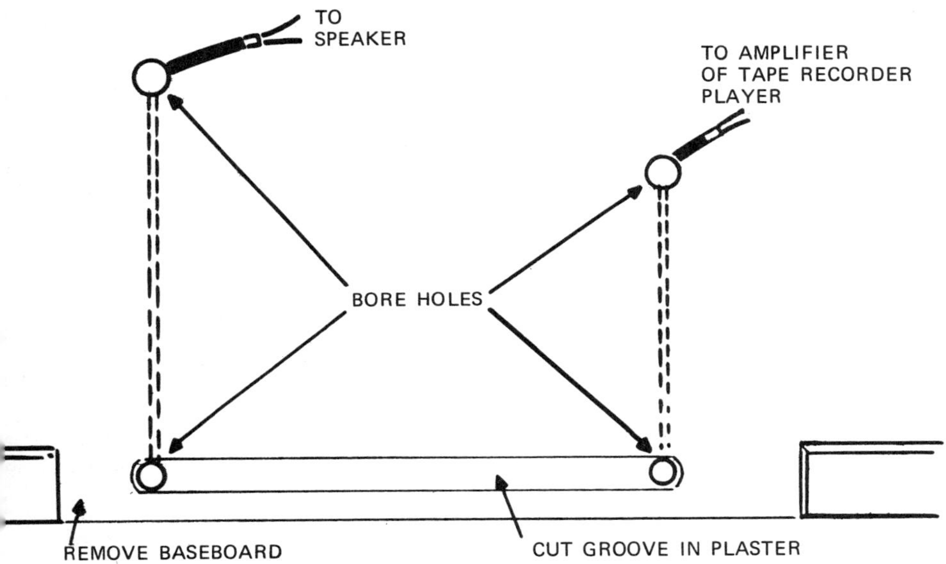

Figure 3-1. How to conceal cables for extension speakers.

To run cables across the room, remove the baseboard and notch a channel in the plaster behind it, running from a spot under the recorder to a spot under the speaker. Bore a hole at both ends. Bore a hole behind the recorder and another hole behind the speaker. Insert a fishwire in the recorder hole and another fishwire in the baseboard level hole beneath it. Connect the two fishwires and pull the cable through. Run the cable in the notched channel, stapling it if necessary. Insert the fishwire through the baseboard at the other end, another fishwire through the hole behind the speaker, connect them and pull the cable through. Replace the baseboard (being careful not to drive a nail through the cable) and plaster the holes.

You can run a cable around door frame headers in the same

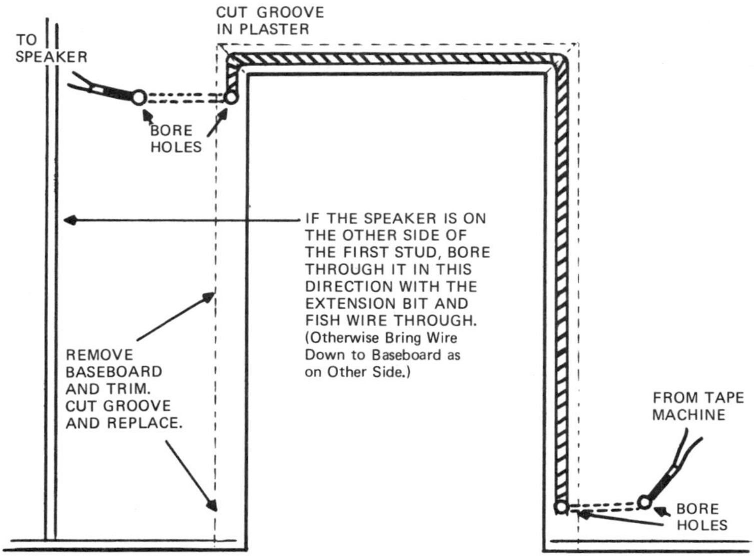

Figure 3-2. How to run a cable around a door.

manner. Remove the baseboard and door trim. You may need to bore a hole with an extension bit through the first upright (see illustration.)

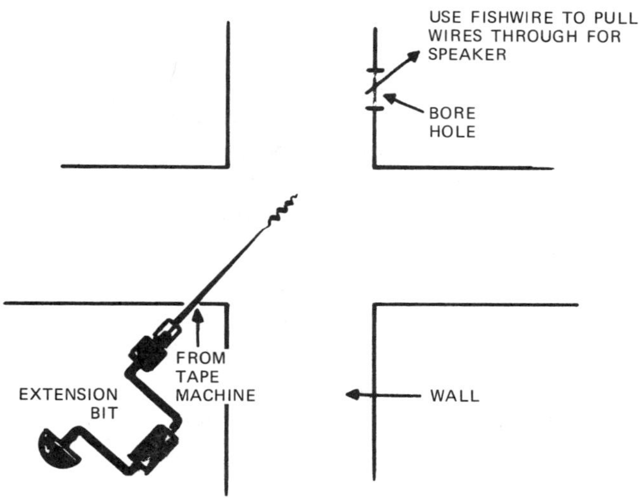

Figure 3-3. How to run a cable from floor to floor.

Running cables from one floor to another will require a long shank bit. Bore through the floor to partition between walls from the floor below. To bore to outer wall of the house, you will need to drill from a diagonal angle (see illustration.) Push a length of fish-wire up through the hole and connect with wires through a hole bored behind the baseboard. Proceed as previously described to run the cable to any part of the room.

You can also tack the cable to the baseboard with electrician's staples driven in with a tack hammer. A cable staple gun (using 1/4 by 3/8 inch staples) makes the job easier and quicker. Use cable the same color as the baseboard or wall (or paint it). You can "wrap up" your wiring job with nylon or twist-lock wire ties, or insert it in flexible vinyl sleeving which comes (in colors) in two-foot lengths. You can even use wire in a contrasting color to add a decorative touch.

Figure 3-4. A typical finished professional installation. (Courtesy of Zenith)

Auto installations make use of standard mounting devices. There are three basic types of installations: under-dash, L brackets, and floor mounting. Your installation needs should be confined to only one of these installations so that you can perfect your procedures for quick servicing. If you use all of them, you will find that combined with the great variety of models of cars, each job will be so different that you will need to spend too much time. Center your work around only one type of installation.

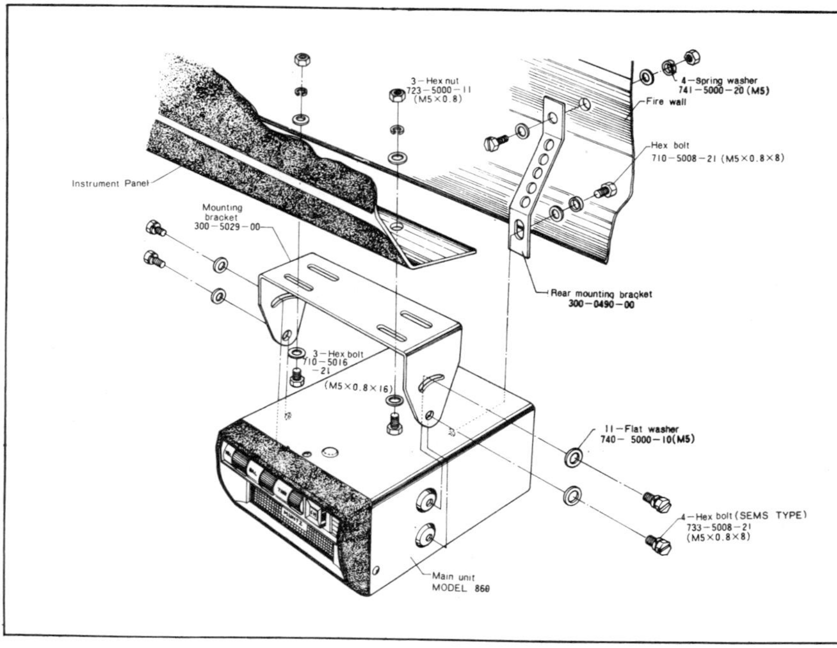

Figure 3-5. A typical auto installation. (Courtesy of Clarion Corporation of America, Los Angeles, California)

The most common installation method seems to be the under-the-dash type. Installations will usually require drilling no more than four holes using the template furnished with the recorder or holder. If you use the same recorder-holder for all installations, you will be able to cut down on the time needed by making a permanent template. Mount the brackets using the necessary hex bolts, bracket washers and hex nuts. Mount the rear of the car tape player to the body of the car by using the rear mounting brackets.

Universal "L" brackets are used in a similar manner; in fact many factories supply a similar bracket with their equipment. However, since it is "universal" (made to fit all) you'll find that it will fit no player *exactly*. This means you'll have to spend valuable time making it fit correctly.

Floor mounts are generally used in sports cars or wherever space is very limited. Installation is simpler in that you'll be working from above, not underneath; however, you'll still be working on a "universal" mounting system, which means nothing will fit precisely.

Speakers may be installed in two basic ways: surface mounted and flush mounted. Surface mounted speakers (no holes to cut) may be positioned under the seat, under the dash, on the rear shelf, on the roof panel, or on the door posts. Installation means merely fastening them down with sheet metal screws.

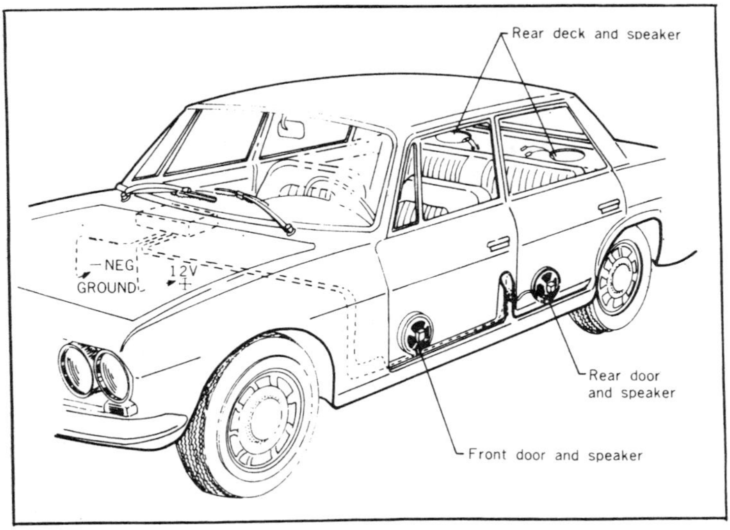

Figure 3-6. Places to mount speakers. (Courtesy of Clarion Corporation of America, Los Angeles, California)

Flush mounted speakers require you to cut holes so that they may be recessed. You can mount them on the rear deck, in a door panel, or in the kick panel. Most domestic cars have a pre-cut opening in the rear deck (usually 6 by 9 inches). Simply remove the opening, install the speaker and cover with the black plastic grille which comes with it.

Most car stereo tape players are designed for use with either two or four speakers—one on each side or two in the front and two in the rear (right and left sides of the car). Run the connecting cables from the equipment to the speakers) concealed under the lower edge of the door openings (see illustration.)

Wire connection is simple. Connect the red lead to the positive (+) terminal of the ignition switch. (Be sure power is supplied to the unit only when the switch is turned "on.") The black lead is connected to the chassis of the car. (Make sure a good electrical connection exists.) Connect the green lead to the plus side of the left speaker; the minus side of the speaker is connected to the chassis ground. Route the gray lead to the plus side of the right speaker and connect the minus side to the chassis ground. Insert a tape and play. (If the unit includes a radio, you will need to connect the car antenna.)

All installations will be variations on these simple procedures. Keep things uncomplicated by using conventional brand name components and by standardizing your methods.

Profit-building ways to install equipment in cars, homes, and businesses

Since you are an electronics technician, you are not going to make much money if you get too involved in non-electronic work; thus, you must keep installation chores to a minimum. If you get too many specialized installation jobs, much of your time will be given over to semiskilled jobs like boring holes, threading screws and rigging up makeshift devices.

There are other alternatives which will still permit you to make money while getting these jobs done. For car player installation, you might be wise to farm the work out. Arrange with a nearby shop specializing in auto radio and tape recorder servicing to have your work done for you at a price that will permit you to make a profit. When your customer brings in his car for installation, call the other shop to have them pick it up. Have the car returned in time for the customer to pick it up in the evening.

Installations in homes and businesses can also be taken care of in this manner. You may find allied businesses willing to take on

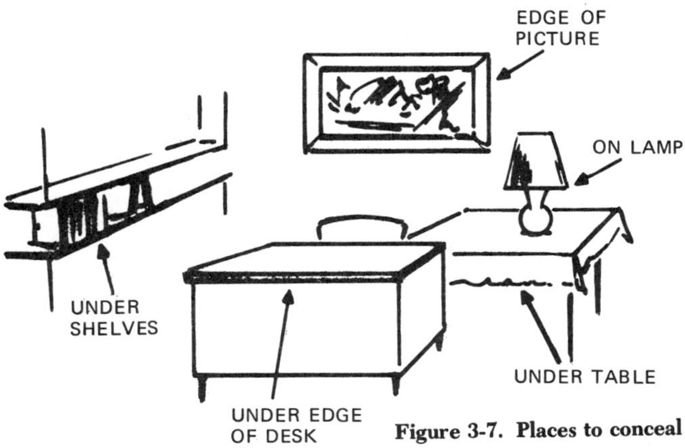

Figure 3-7. Places to conceal microphones.

such installations on a flat rate basis (installers of fire alarm and burglar alarm equipment sometimes do such work during slack periods).

Work installations in with other service calls, or into periods in which you are not busy, and you will find you can increase your profits.

Hiding secret recorders for surveillance

You can find possibilities for profits in the burgeoning security business. Because crime is increasing, private detective agencies often have requests from companies for installation of recording equipment for surveillance. This equipment ranges from sophisticated voice-activated recorders (which begin working when someone speaks) to simple recorders (with built-in condenser mikes) which are left running in rooms under surveillance. A microphone (as small as 13/32 inch—.1 ounce) can be hidden in the room under a desk, behind a chair, or on a lighting fixture, and connected with a long cable to a recorder in another room. More complex devices may use a miniature AM or FM transmitter whose signal is picked up by a receiver in a nearby vehicle or room and recorded on tape. Although little use has been made of it at present, a parabolic reflector (which looks like a bowl and is designed to concentrate sound to a focal point—a mike in its center) can be used some distance away from the sound source merely by pointing it at the speaker, thus recording a conversation.

Obviously, such activities do not appeal to everyone. If security interests you, however, explore the possibilities of supplying equipment and making installations for nearby firms.

Chapter 4

Open-Reel Transport Mechanisms: Service and Repair

Mechanisms of tape recorder/players are often complex but are not as difficult to service and repair as record changers.

Basic methods used to transport tape

Open-reel machines move tape from one reel to another by one of the following methods: (a) a capstan (a driven spindle or shaft which rotates against the tape, pressing it against a rubber wheel) drives the tape through the machine at a steady speed while a take-up reel pulls the tape—both run by a single motor (b) three motors are used—one drives the capstan, one drives the take-up reel, and one (at half voltage) maintains an electrical drag or tension against the tape movement from the feeding reel. A simple system of belts or drive wheels is used to change speed or to reverse (see illustration).

Basic modes of operation are *play* and *record*, *fast forward*, and *rewind*. Some recorders also provide for *pauses*. Localize your service problem so you can trace it in one (or more) of these modes.

In the modes of operation you will find malfunctions of various parts quickly if you proceed systematically. Some parts will be replaced frequently; some will seldom, if ever, need replacement.

Motors used in recorders are of three types. Most inexpensive and medium-priced recorders use single-phase shaded-pole induction motors (in which the speed is controlled by voltage changes) which generally run at a speed close to 1800 rpm. More expensive recorders use either a salient pole synchronous motor or a hys-

Figure 4-1. A capstan.

teresis synchronous motor (in which the speed is controlled by current frequency—like a clock). The third type is the dc battery motor (in which the speed is governor controlled by varied voltage input). Motor failures account for only a small proportion of malfunctions.

The *capstan* pulls the tape evenly through the machine and minimizes speed variations caused by fluctuations in the motor drive and tape thickness. Early capstans were rubber; more modern ones are steel. Generally you will need only to clean the capstan; sometimes you will need to replace it (if it is obviously damaged).

Figure 4-2. A three-speed drive system.

A *flywheel* usually is attached to the bottom of the capstan shaft and is driven by either a belt or puck (a drive wheel). Belt drives may break or slip or stretch. The puck may slip from oil or grease or may become flat in one spot (from being stopped in an operational mode). Many service jobs call for no more than the replacement of one of these belts (look for rubber "sawdust" which indicates a damaged belt) or pucks (look for flat or smooth edges).

Rewind/fast-forward drive systems generally use a system of belts or drive wheels. Malfunctions are most often solved by replacing or cleaning these rubber parts.

The *take-up reel* uses a variable speed drive (its rpm changes from fast to slow as more tape is wound on the reel). This is done with a slipping belt, a stalled induction motor, or a slipping clutch. Repairs are infrequent, but usually require replacement of the part.

Braking mechanisms usually consist of a pad or wheel which rubs against a metal disc on the rim shaft. The reel from which the tape is coming is stopped first to avoid spillage. Adjustments are fairly common.

Most mechanical repairs will consist of cleaning and adjusting. Replacements are common for rubber parts (belts and drive wheels), but less common for other parts (motors and clutches).

A simple check-list for correcting operational difficulties

When the customer explains the problem, listen carefully, but beware—many times the malfunction is not the fault of the equipment. Check it out yourself.

Use a quality reel of tape in your check (the fault may be in a defective tape reel), use your own dependable electric cord and socket (a damaged cord or a dead socket may be the cause), and check to see that the equipment is placed in the mode called for (if the customer forgets to press the right button, he won't get the results expected).

Luckily, most true mechanical malfunctions are as easy to spot as a broken belt, and repairs are simple once you pinpoint the trouble. You will then need to clean, lubricate, adjust or replace parts. Use the following list as a handy guide to make quick and easy profit-building repairs. Repair the trouble by doing *one* of the following:

1. *If the motor doesn't run*
   a. replace the on-off switch (if it is defective).
   b. replace the micro-switch (if the on-off switch is o.k. but the micro-switch is malfunctioning).
   c. replace the capacitor
   d. replace the motor

Figure 4-3. Be careful not to bend the blade on the motor!

2. *If the capstan doesn't turn*
   a. replace the broken drive belt.
   b. tighten the loose screw on the pulley (if any)
   c. replace motor
3. *If the tape starts too slowly*
   a. clean, adjust or replace drive belt.
   b. correct or replace bent motor shaft.
4. *If the tape speed is too fast*
   a. replace the motor governor.
5. *If the tape speed is too slow*
   a. clean or replace drive wheel.
   b. clean or replace flywheel.
   c. clean or replace capstan.
   d. replace thrust cap of capstan housing.
   e. stop reel from slipping.
   f. stop pressure roller from slipping.
   g. stop belts from slipping.
   h. lubricate reel shafts or rotor shafts.
   i. clean greasy roller or idler wheel.
   j. replace motor.

Figure 4-4. A flywheel and drive system.

6. *If the take-up reel doesn't move*
   a. clean and adjust pressure roller.
   b. stop take-up reel from slipping.
   c. stop drive belt from slipping.
   d. correct and adjust brake action.
   e. adjust weak rewind torque (check specifications—usually about 600 to 1000 grams).

7. *If the tape breaks* (cuts, tears, winds around capstan)
   a. repair or replace capstan housing (if it is damaged, dirty or worn).
   b. straighten or replace the pressure roller plate (if it is damaged).
   c. align heads (see Chapter 7).

8. *If the tape spills*
   a. replace the broken drive belt.
   b. adjust the brake (usually no more than 1/64 inch clearance is required).
   c. replace the defective clutch unit.

9. *If the fast-forward/rewind is inoperative*
   a. clean or replace the fast-forward/rewind belt.

Figure 4-5. A motor and belt system.

b. repair or replace the inoperative leaf switch.
c. correct tension on the fast-forward/rewind spring (check against specifications—usually 600 to 1000 grams).
d. stop the roller from slipping.
e. adjust the brake plate so it doesn't touch (usually 1/64 inch clearance is required).
f. replace the damaged pinch roller.
g. adjust the reel back tension.
h. replace the bent reel shaft.

10. *If the automatic-stop feature malfunctions*
    a. repair or replace the inoperative lock spring.
    b. clean or replace the pinch roller.
    c. correct (or replace) control circuit switch.
    d. replace faulty solenoid.

Stopping wow and flutter

Wow (periodic variations in tape speed causing distortions in pitch) and *flutter* (short rapid variations in tape speed causing distortion in pitch and volume) are such common servicing problems that a separate section needs to be devoted to them. You can check for wow and flutter by using a commercially made wow-and-flutter tape (with a recorded flutter of less than .03 percent), by using a wow-and-flutter meter (excellent ones are manufactured by Ferrograph, Manke or McAdam Electronics, Inc.), or simply by listening to the recorder playing a piano selection.

The most common causes and remedies in open-reel recorders are as follows:

a. use a different reel.
b. clean or replace the capstan shaft assembly.
c. adjust the pinch roller and spring.
d. adjust improper take-up torque (check specifications).
e. clean and adjust dirty or nonaligned heads.
f. replace the pressure roller.
g. replace the defective drive belt.
h. lubricate the flywheel or bearing (sparingly).
i. correct the thrust gap.
j. replace felt pads.
k. clean greasy idler drive wheel, flywheel or pulley.
l. adjust tension of idler.
m. stop motor pulley from slipping.
n. replace motor.

How to perk up worn equipment

Some of your customers have old equipment of which they are fond, and they'll want you to pep it up or restore it to near its original condition. Sometimes this will be too expensive to make it worthwhile and you'll simply have to sell them on the idea of getting a new machine. As rule of thumb, if you estimate the repairs will cost more than half of what the machine is worth on the open market, it won't be worth it.

If the equipment is worth repairing, the first step is a thorough cleaning, inside and out. Wipe dirt off with a dry cloth and lubricate necessary parts sparingly; then replace all the rubber parts possible.

This step is one that may be more costly than it is worth, because it may be impossible to get all the needed replacement parts. Also, a machine seems to wear proportionally, and sometimes replacing all of one item (like the rubber) may cause stresses and strains which would not have occurred if the equipment had been left alone. Most servicemen thus stick to the motto of replacing only what needs replacing.

Restore the tension to original specifications on the pinch roller, the fast-forward and rewind rollers, the idler drive wheel and the pressure pad. Sometimes the springs supplying the required tension weaken; replace them if necessary. Use a tension gauge to make the measurements and check them against specification sheets and service manuals available for specific models.

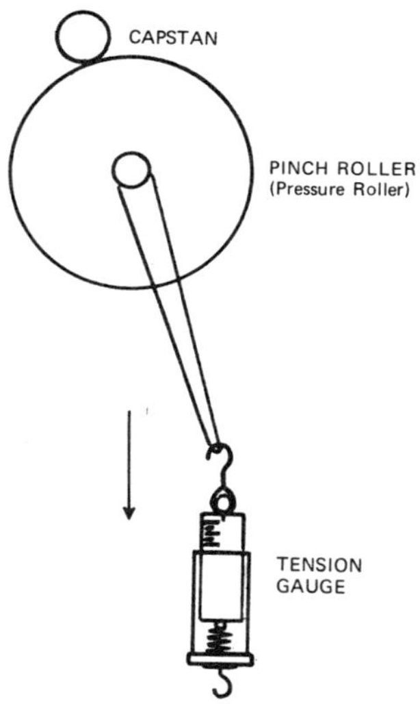

Figure 4-6. How to measure pinch roller (pressure roller) tension.

To check the *pinch roller* tension, connect the tension gauge to the pinch roller shaft (you might need to use a piece of string), press the mode switch, and pull the gauge back until the pinch roller disengages. Check the reading: the average is usually between 1000 and 1500 grams. The tension is changed by altering or changing the spring.

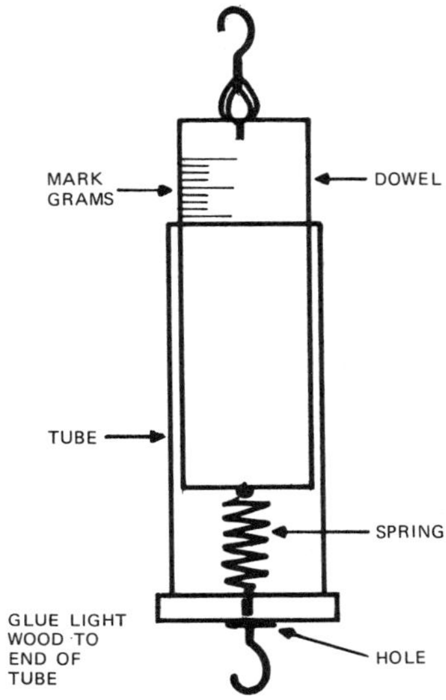

Figure 4-7. A gram tension gauge that you can make yourself.

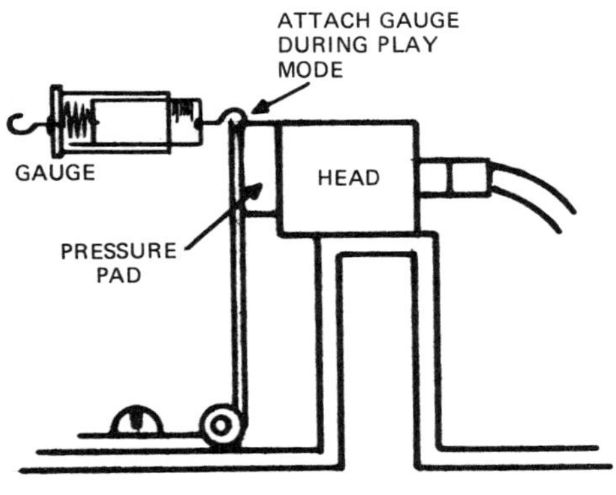

Figure 4-8. How to measure pressure pad tension.

Check the *fast-forward* and the *rewind rollers* by attaching the tension gauge to the shaft of either. Depress the appropriate mode switch and check the reading against the specifications. The average for open-reel recorders ranges from 500 to 1000 grams. Adjust or replace the spring.

The *idler drive wheel* tension is checked by attaching the tension gauge to the idler wheel shaft and putting the recorder into the play mode. Pull in the same direction that the spring pulls the wheel until it disengages. Read the results on the gauge and check it against specifications. The range is usually between 220 and 280 grams for open reel recorders. Adjust or change the spring.

Check *pressure pad* tension by attaching the gauge to the top of the pad. Place the equipment in the play mode and take a reading. Open-reel recorders usually measure in the range between 20 and 40 grams. Adjust or replace as necessary.

Tune up the equipment further by checking the electronic components as described in Chapters 7, 8 and 9. Reassemble the recorder in its case, polish it up a bit and watch your customer happily reach for his checkbook to repay you for the good job you've done in making his recorder function like new!

REALISTIC®

Service Manual

14-973

999B
3-SPEED 3-HEAD
STEREO TAPE DECK

Catalog Number: 14-973

CUSTOM MANUFACTURED FOR RADIO SHACK A TANDY CORPORATION COMPANY

SPECIFICATIONS

POWER SUPPLY	AC 120 V, 60 Hz
POWER CONSUMPTION	25 W
RECORDING COMPENSATION SYSTEM	NAB
RECORDING SYSTEM	AC bias
ERASING SYSTEM	AC erase
BIAS OSCILLATOR FREQUENCY	95 kHz ± 7 kHz
INPUT IMPEDANCE	
MIC JACK	10 kohm
AUX IN JACK	300 kohm
OUTPUT IMPEDANCE	
PREAMP OUT JACK	10 kohm or less
HEADPHONE JACK	8 ohm
INPUT LEVEL	
MIC JACK	280 μV ± 4 dB
AUX IN JACK	77.5 mV ± 3 dB
OUTPUT LEVEL	
PREAMP OUT JACK	775 mV ± 2 dB
TAPE REEL	Up to 7″ reel max.
RECORDING OR PLAYING TIME	
STEREO(7″, 35μ tape)	1.5 hr at 7½ ips 3 hr at 3¾ ips 6 hr at 1⅞ ips
MONO(7″, 35μ tape)	3 hr at 7½ ips 6 hr at 3¾ ips 12 hr at 1⅞ ips
OPERATION SYSTEM	One-hand knob system
SEMICONDUCTOR COMPLEMENT	
TRANSISTORS	2SC458D(LG) × 6, 2SC458D × 8 2SC1213 × 4, 2SB370 × 1
DIODES	IN34A × 2, V06A × 4
TRACK SYSTEM	4-track stereo
DIMENSIONS	7⅜″(H)×16″(W)×13¼″(D)
WEIGHT	22 lbs.

DESCRIPTION	NOMINAL	LIMIT
FREQUENCY RESPONSE		
7½ ips	40 - 20,000 Hz	50 - 15,000Hz ± 4 dB
3¾ ips	40 - 10,000 Hz	50 - 7,500Hz ± 6 dB
1⅞ ips	40 - 5,000 Hz	50 - 5,000Hz ± 6 dB
SIGNAL TO NOISE RATIO	45 dB	40 dB
ERASE RATIO	65 dB	56 dB
CROSS TALK	65 dB	55 dB
TAPE SPEED		
7½ ips		±2 %
3¾ ips		±2 %
1⅞ ips		±3 %
WOW & FLUTTER		
7½ ips	0.2%RMS	0.25%RMS
3¾ ips	0.25%RMS	0.32% RMS
1⅞ ips	0.47%RMS	0.60% RMS

BLOCK DIAGRAM

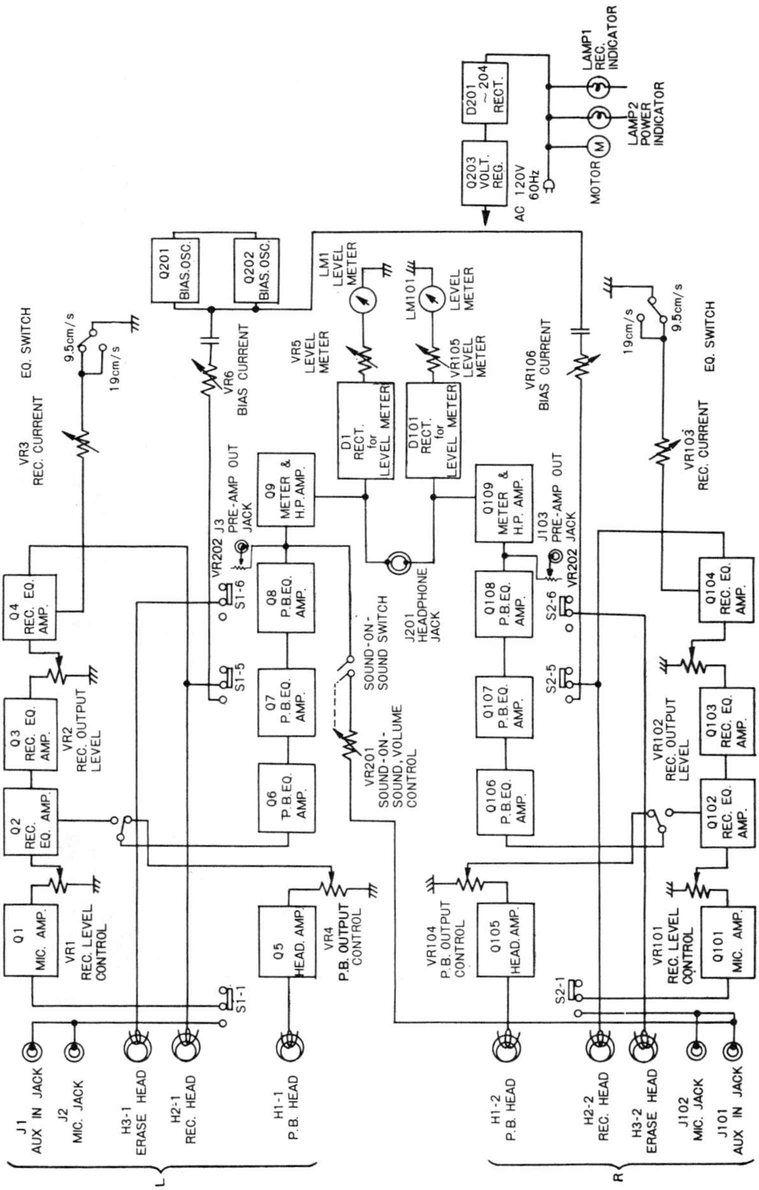

FIGURE 1

DISASSEMBLY

When inspecting, repairing or lubricating, disassemble the recorder in the following manner.

1. **Removal of rear cover**

   Remove four screws holding rear cover and four screws holding chassis shown in Figure 2.

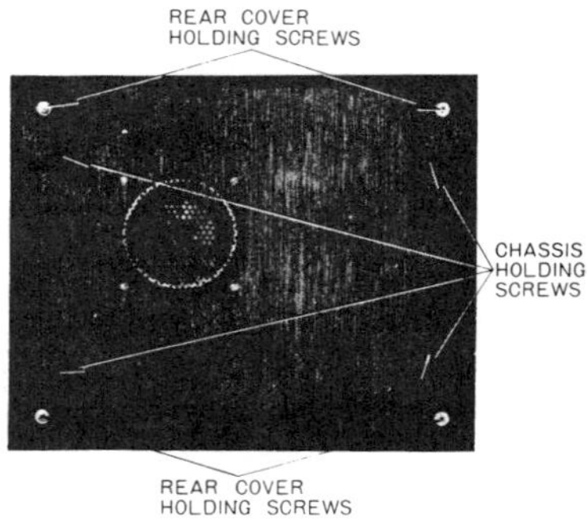

FIGURE 2

2. **Removal of cabinet**

   Remove four screws holding chassis and one screw holding power cord shown in Figure 3.

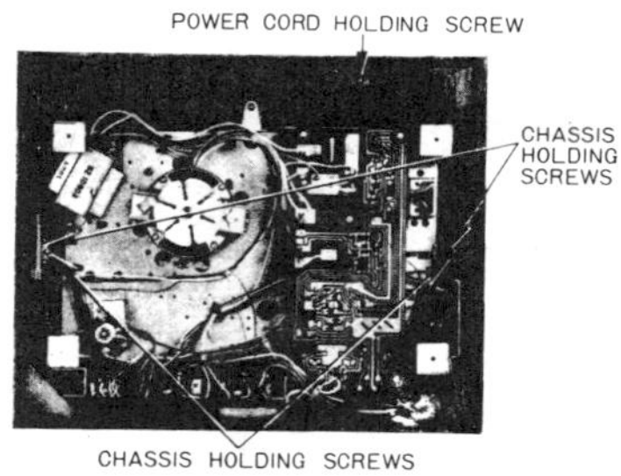

FIGURE 3

3. **Removal of front panel**

   1) Remove one screw holding head cover shown in Figure 4.

FIGURE 4

2) Remove eight screws holding front panel shown in Figures 5 and 6.

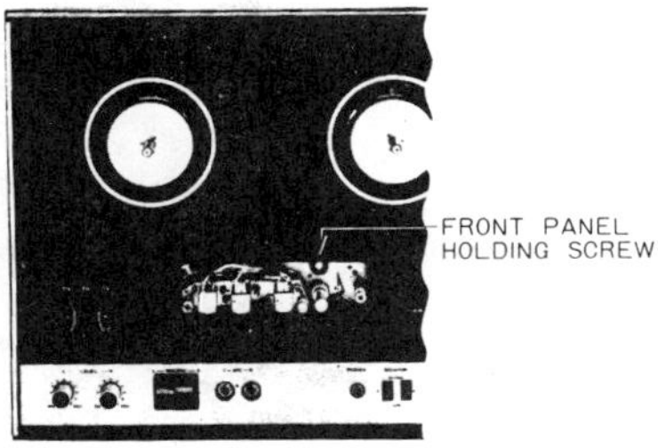

FIGURE 5

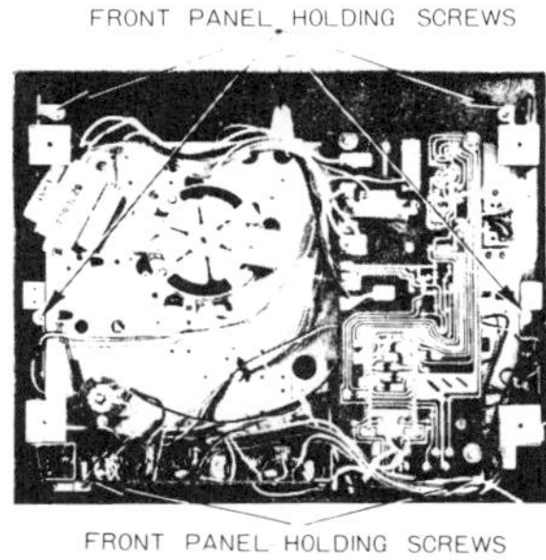

FIGURE 6

4. Removal of circuit board

Note: Take care with the wires connecting parts of circuit board.

1) Circuit board for amplifier block: Remove five screws holding circuit board shown in Figure 7

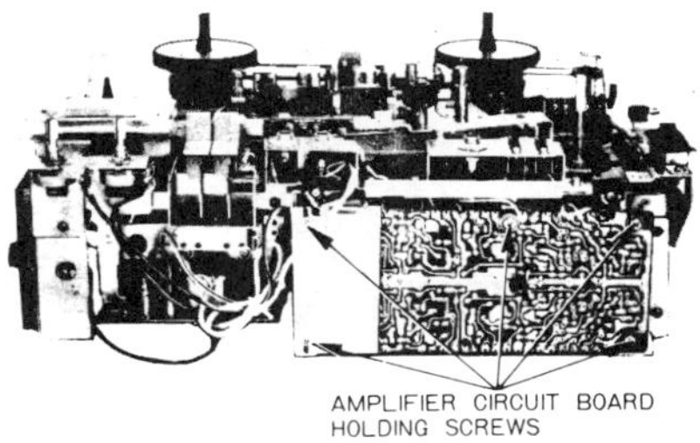

FIGURE 7

2) Circuit board for switching block: Remove five screws holding circuit board shown in Figure 8.

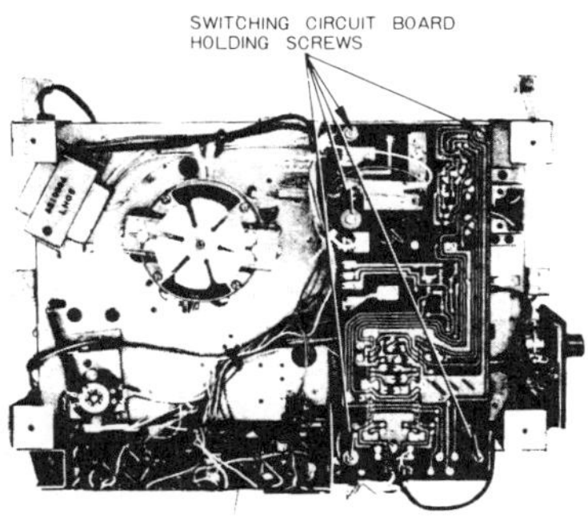

FIGURE 8

LUBRICATION

One or two drops of machine oil may be applied to the points Ⓜ (rotating parts).
Grease may be applied to the sliding surfaces Ⓖ as shown in Figure 9.

Ⓜ : Machine oil
Ⓖ : Grease

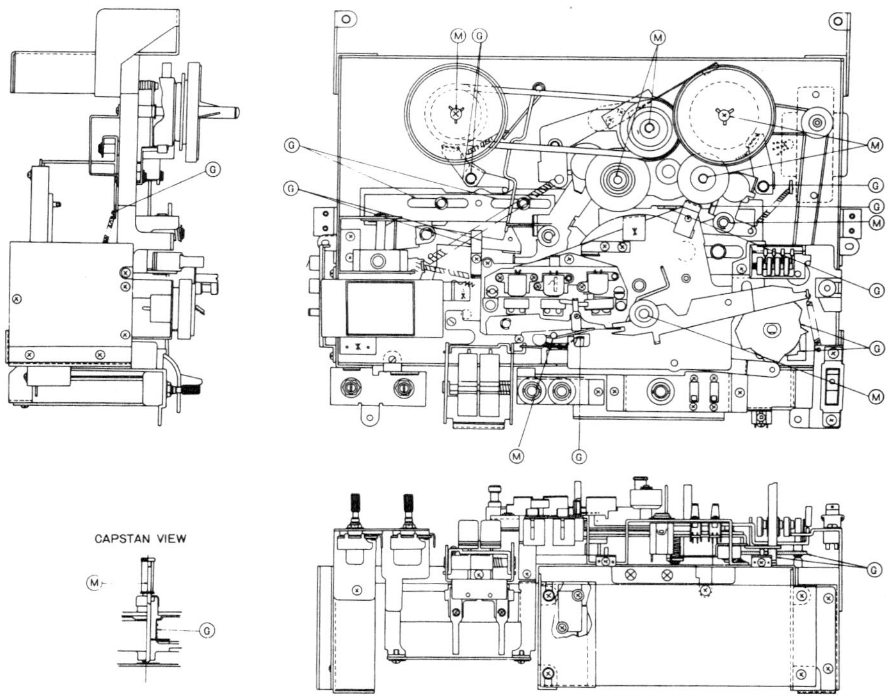

ADJUSTMENT

1. Electrical Adjustments

Make sure that all heads are demagnetized and cleaned before making any adjustments.

1-1 Height adjustment of Record and Playback heads.

1) Use a transparent tape and run it through the tape guides.

2) Adjust the tape guide so that the upper edge of the tape just matches the upper end of the track core at the top of the head.

1-2 Azimuth adjustment of Playback head

1) Playback a 15 kHz signal using AMPEX test tape #01-3132104 and adjust Azimuth screw for maximum output.

2) When the output of R and L channels is different, adjust Azimuth for best compromise between channels.

3) Chech pressure pad adjustment. Proper adjustment should result in maximum 15 kHz response.

1-3 Adjustment of L1 and L101

1) Make the following settings: depress the RECORD buttons, set Selector lever to STOP, set Record LEVEL controls fully clockwise and set VR3 and VR103 to maximum.

2) Apply a 18kHz (−90dB) signal to MIC jack and monitor both sides of the head.

3) Adjust L1 and L101 for maximum and equal reading on VTVM.

1-4 Gain adjustment of playback amplifier

1) Playback a 1 kHz signal using AMPEX test tape #01-31321-04.

2) Adjust VR4 and VR104 so that playback amplifier output is −10 dB (245 mV).

1-5 Azimuth adjustment of Record head

1) Apply a 15kHz signal to MIC jack at a level which will produce −20 dB output.

2) Place unit in Record operation and set TAPE SPEED to 7½ ips.

3) Monitor Playback output by turning Monitor switch to TAPE (both channels).

4) Adjust Azimuth for maximum playback output of recorded signal.

5) If maximum points of L and R channel is different, adjust for best compromise between channels.

6) Again check pressure pad adjustment for max high frequency response.

1-6 Bias leakage adjustment

1) Set unit in recording operation and turn PREAMP OUT control at maximum.

2) Monitor the voltage at L2, L102 and R23, R123 and adjust L2 and L102 for minimum.

1-7 Bias adjustment

1) Make the following settings: depress the RECORD buttons and set TAPE SPEED selector to 7½ ips.

2) Apply a 1 kHz signal (level to produce −20 dB) to MIC jack and adjust bias current for max output.

3) Apply a 10 kHz signal (same input level as above)to MIC jack and check output.

4) Set VR 6 (VR 106) so you obtain equal output at both frequencies with a given input level.

1-8 Adjustment for recording current

1) Make the following settings: depress RECORD buttons and set TAPE SPEED selector to 7½ips.

2) Apply a 18 kHz signal to MIC jack and adjust VR3 and VR103 so that playback output is the same as that for 1 kHz.

3) If equal output cannot be obtained, adjust close as possible.

1-9 LEVEL meter adjustment

1) Make the following settings: depress RECORD buttons and Monitor switch to SOURCE.

2) Apply a 1 kHz to MIC jack and adjust it so that output level of LINE OUT is 0 dB.

3) Adjust VR5 and VR105 so that the pointer of LEVEL meter indicates 0 VU.

4) Set the Monitor switch to TAPE and adjust VR4 and VR104 so that playback output reads 0 dB on LEVEL meter.

1-10 Playback inductive hum adjustment (L3 and L103)

1) Set unit in playback operation without tape.

2) Loosen adjusting screw to rotate coil, and tighten the screw at the point where inductive hum is minimum, monitor with VTVM.

3) Adjust core to where inductive hum is minimum on VTVM.

2. Inspection of Mechanical Parts

2-1 Pressure

These measurements are to be made with the tape recorder mounted horizontally.

Item	Parts	Pressure	Method of Measurement
1	Pressure roller	900-1,300 gr	Measured with a 2 kg bar tension gauge. Turn POWER to ON and Selector lever to PLAY. Note the amount of pressure when Pressure Roller stops turning, while the Pressure Roller shaft is being pulled away from the capstan.
2	Fast forward idler	150^{+50}_{-30} gr	Measured with a 200 gr sector tension gauge at the Fast forward idler shaft. Turn POWER to ON and Selector lever to PLAY. Note the amount of pressure required to make the Fast forward idler simultaneously separate from the Motor Pulley and take-up reel Table (without reel).
3	Play idler	200±40 gr (at 7½ ips) Reference 175±40 gr (at 3¾ ips) 150±40 gr (at 1⅞ ips)	Measured with a 300 gr sector tension gauge at the Play idler holder. Turn POWER to ON and Selector lever to PLAY. Note the amount of pressure required to make the Play idler simultaneously separate from the Motor Pulley and flywheel (without reel).
4	Rewind idler	450±50 gr	Measured with a 700 gr sector tension gauge. Turn POWER to ON and Selector lever to REWIND. Note the amount of pressure when the Rewind idler stops turning while the Rewind idler is moving away from the Motor Pulley along a line between the shaft of Rewind idler and Motor Pulley.
5	Pressure pad	25-35 gr	Measured with a 50 gr sector tension gauge at the top of pad. Set the Selector lever to PLAY. Note the amount of pressure when the erasing and REC/PB pad respectively are separated from the side of the magnetic head.

2-2 Torque

Item	Parts	Measured Value	Method of Measurement
1	Take-up force	50 - 90 gr	Measured horizontally with a 100 gr sector tension gauge with the set in vertical position. In ON (power) and PLAY, the tangential force is measured at a radius of 30mm with a 7″ empty reel.
2	Torque on take-up friction coupling	200 - 300 gr. (upper ref.)	Measured horizontally with a 500 gr sector tension gauge with the set in vertical position. In ON (power) and F. FWD, the tangential force is measured at a radius of 30mm with a 7″ empty reel.
3	Torque on supply friction coupling	200 - 300 gr. (upper ref.)	Measured horizontally with a 500 gr. sector tension gauge with the set in vertical position. In ON (power) and REWIND, the tangential force is measured at a radius of 30mm with a 7″ empty reel.
4	Back tension at take-up side	15 - 35 gr	Measured horizontally with a 50 gr sector tension gauge with the set in vertical position. In REWIND, the tangential force is measured at a radius of 30 mm with a 7″ empty reel.
5	Back tension at supply side	15 - 35 gr	Measured horizontally with a 50 gr sector tension gauge with the set in vertical position. In PLAY or F. FWD, the tangential force is measured at a radius of 30mm with a 7″ empty reel.
6	Working force of knob	a) Recording knob 3.2 kg (max.)	The value at which the knob is locked is measured while the tip of the knob (hook of jig) is pulled with a 5 kg bar tension gauge.
		b) Speed selector knob 2.8 kg (max.)	The value at which speed is changed by one step is measured while the tip of the knob (hook of jig) is pulled with a 3 kg bar tension gauge.
		c) Function knob 14 kg-cm (max.	A jig (l = 20cm) is put in the change shaft, and the value at which the cam is switched by one step is measured while the shaft is pulled with a 1 kg bar tension gauge at the point l = 20-cm. In this measurement the value for REWIND↔STOP-PLAY↔PAUSE↔F.FWD respectively shall be the same and less than 700 gr.

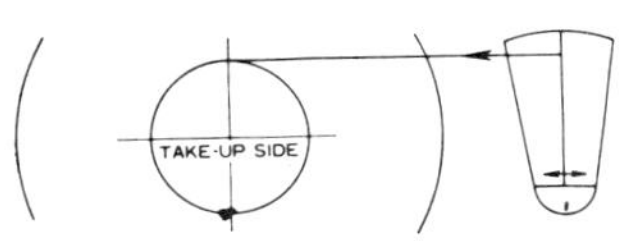

1. For the measurements of items (1) and (2), measure while winding slowly in the direction of the arrow.
2. For the measurement of item (4), measure while pulling gently in the opposite direction.

FIGURE 10

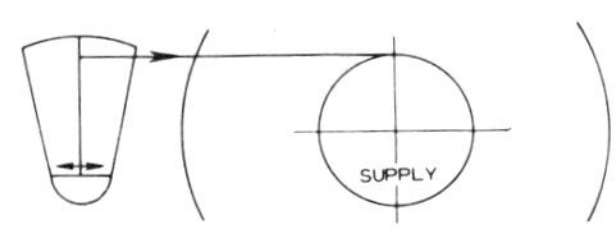

1. For the measurements of item (3), measure while winding slowly in the direction of the arrow.
2. For the measurement of item (5) measure while pulling gently in the opposite direction.

FIGURE 11

3. Mechanical Adjustment

3-1 Motor Pulley (See Figure 12)

The motor pulley is adjusted to the dimensions shown in Figure 12 when the Play idler is positioned for a speed of 1⅞ ips (4.75cm/s), with the set at horizontal position and in PLAY.

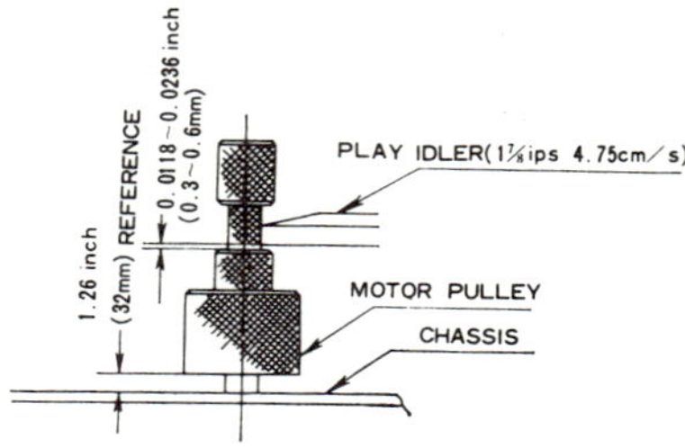

FIGURE 12

3-2 Shut-off Rod and Microswitch (See Figure 13)

With the set at horizontal position and in PLAY, the microswitch holder is held by 3mmφ × 6mm pan-head screws so that the microswitch is switched from OFF to ON at b = 7mm when the shut-off rod is shifted in the Y direction.

At STOP, the dimension (a) is to be more than 1mm. When the set is at REWIND, PLAY or F. FWD and without a tape, the microswitch shall always be OFF.

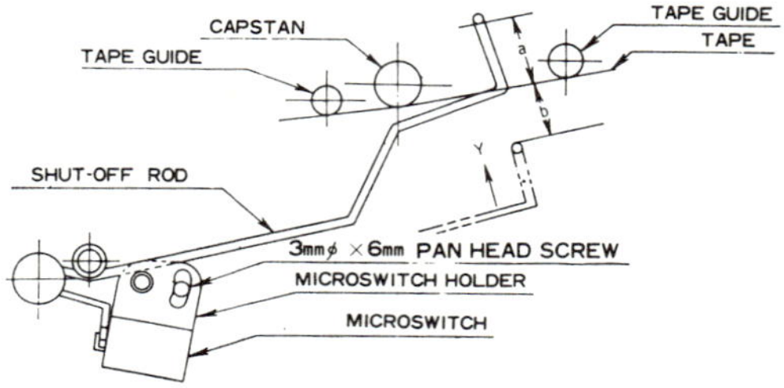

FIGURE 13

3-3 Brake Rubber and Supply or Take-up Reel Table (See Figure 14)

The timing of braking is to be adjusted either at the supply or take-up side by bending the brake lever in the direction of the arrow. Except at STOP or PAUSE, however, the gap between the takeup or supply reel table and brake rubber shall be more than 0.5 mm.

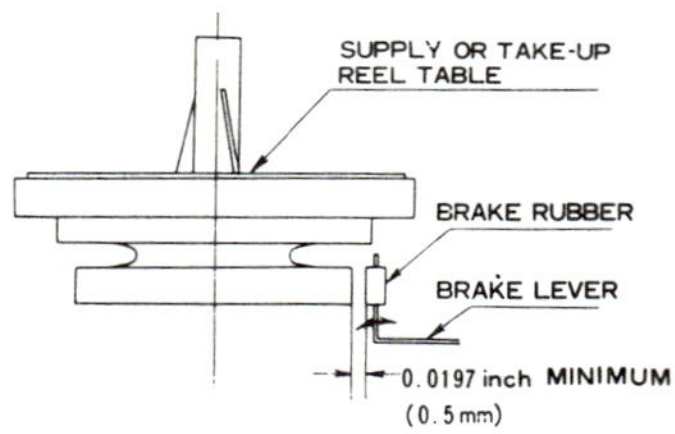

FIGURE 14

3-4 Pressure Roller and Capstan (See Figure 15)
With the set at STOP, the gap between the capstan and Pressure roller (a) in Figure shall be more than 7 mm.

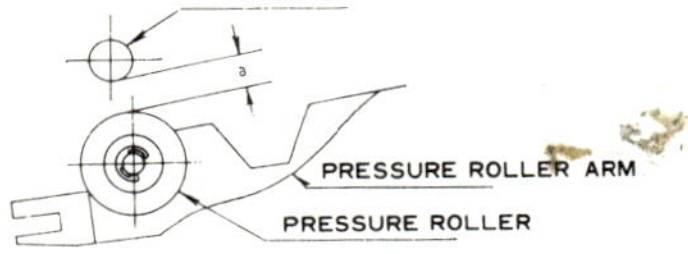

FIGURE 15

3-5 Head Cleaning
Periodically clean the tape heads with standard tape head cleaner and cotton swabs or head cleaning tapes. Clean the heads, tape guides, and capstan shaft with care.

SEMICONDUCTOR LEAD IDENTIFICATION

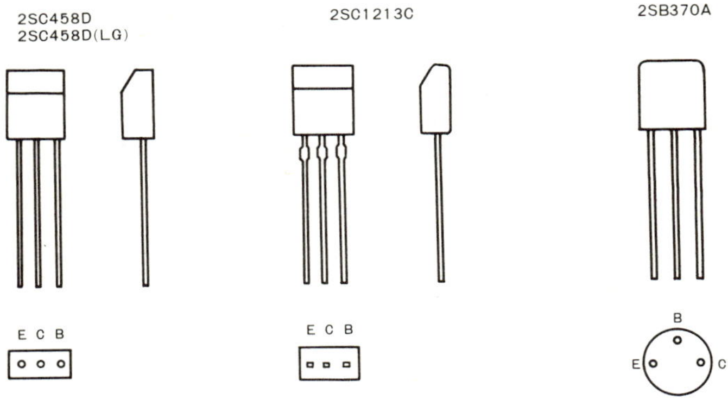

SCHEMATIC DIAGRAM

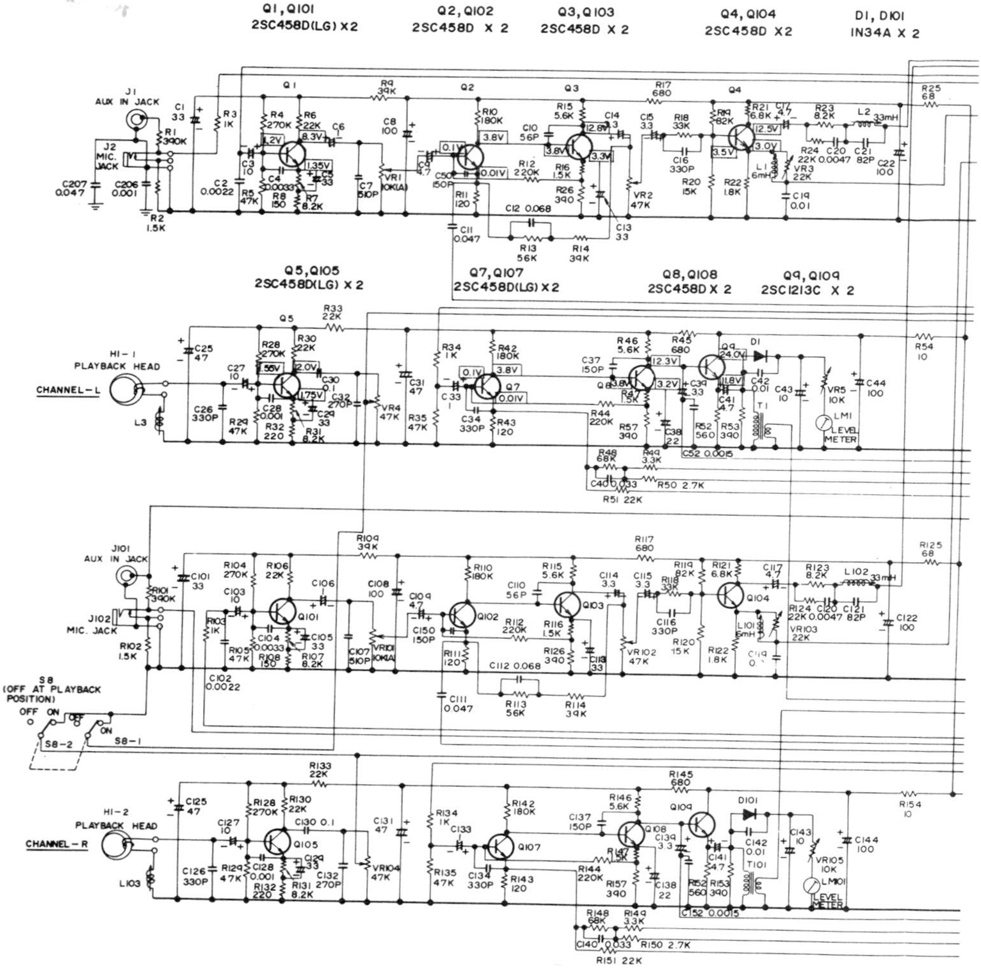

S 1 ; RECORD SWITCH (CHANNEL-L)
S 2 ; RECORD SWITCH (CHANNEL-R)
S 3 ; COMPENSATION SWITCH
S 7 ; MUTING SWITCH
S 8 ; MUTING SWITCH

S 9 ; MONITOR SWITCH (CHANNEL-L)
S10 ; MONITOR SWITCH (CHANNEL-R)
S11 ; SHUT-OFF SWITCH
S12 ; POWER SWITCH

VR1,101 ; RECORD LEVEL CONTROL
VR2,102 ; RECORD OUTPUT LEVEL CONTROL
VR3,103 ; RECORD CURRENT CONTROL
VR4,104 ; PLAYBACK AMP. OUTPUT CONTROL
VR5,105 ; LEVEL METER CONTROL
VR6,106 ; BIAS CURRENT CONTROL
VR201 ; SOUND-ON-SOUND VOLUME CONTROL
VR202 ; PREAMP OUT LEVEL CONTROL

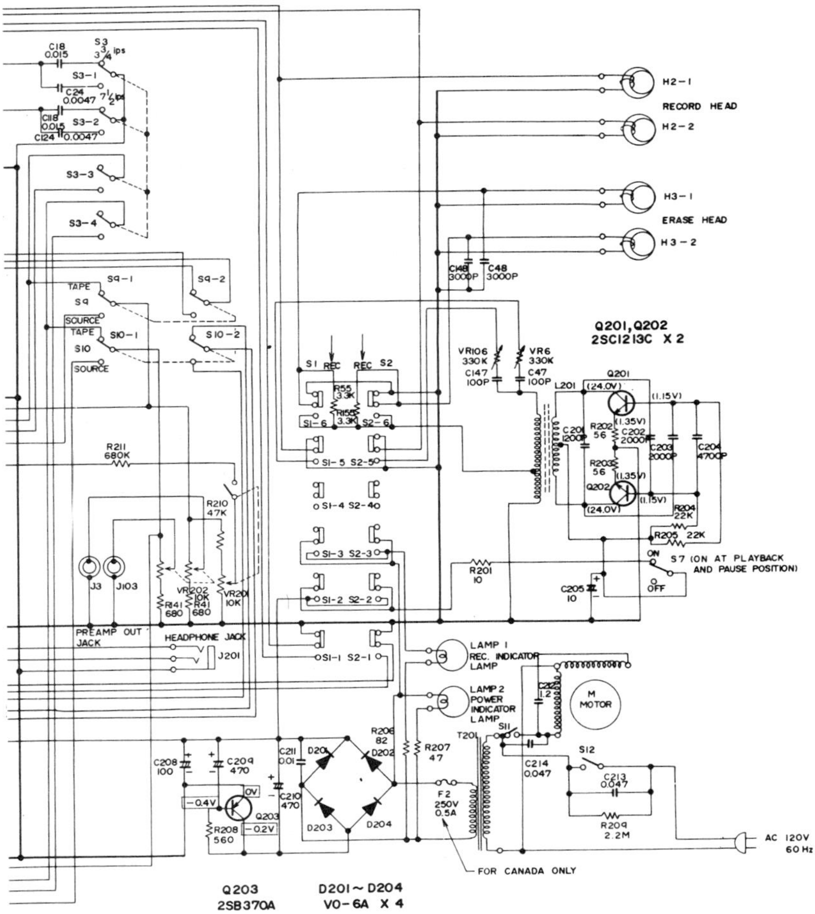

NOTE : 1. ALL RESISTANCE VALUES IN OHMS. K = 1000
2. ALL CAPACITANCE VALUES IN μF EXCEPT "P" NOTED. P = $\mu\mu$F = PF.
3. VOLTAGE MEASURED WITH VOLTMETER FROM ⊖ SIDE.
LEVEL CONTROLS AT MINIMUM AND NO SIGNAL.
ARE SHOWN IN RECORD MODE.

CIRCUIT BOARD DIAGRAM (TOP VIEW)

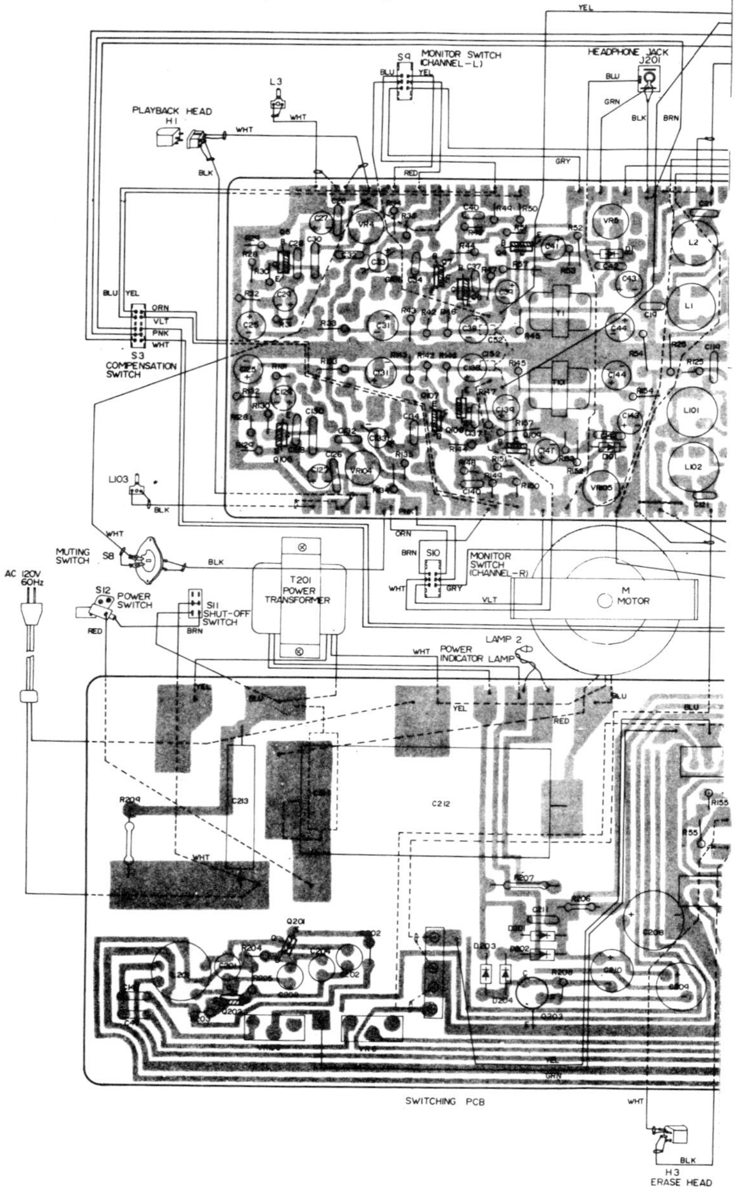

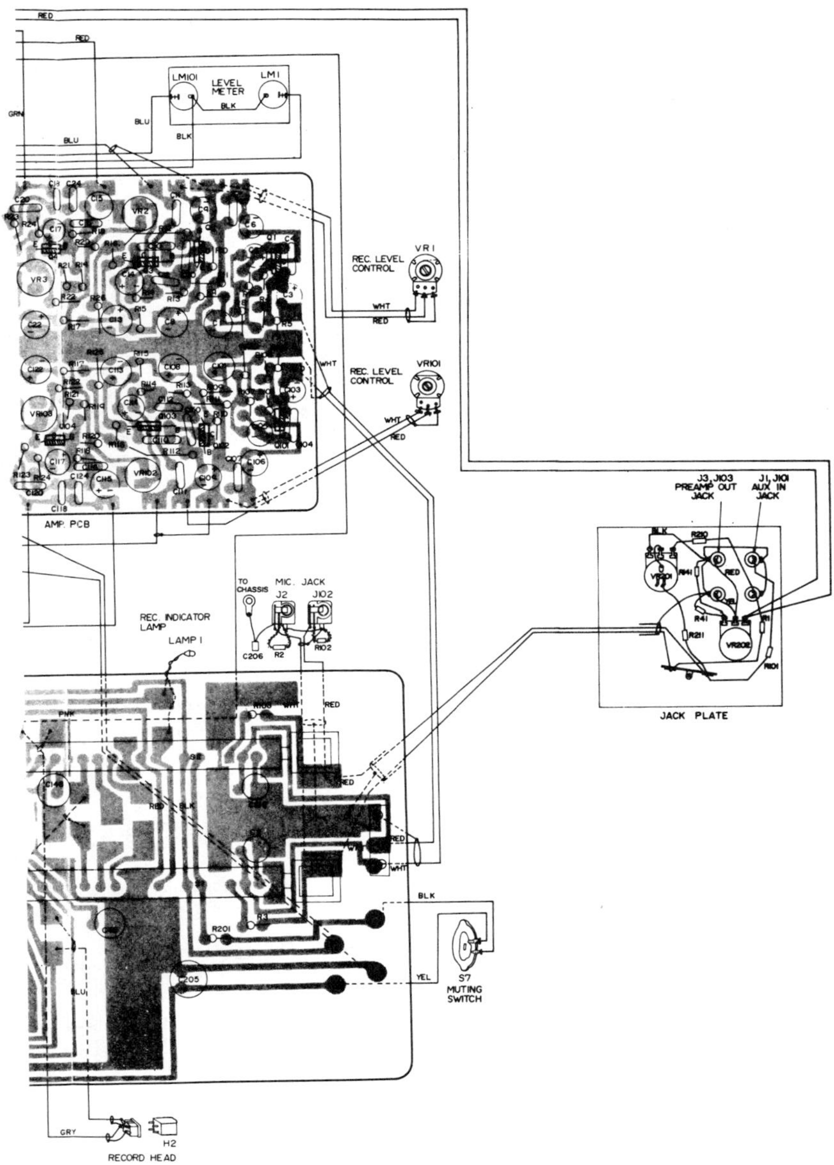
LEVEL METER
LM101
LM1
REC. LEVEL CONTROL
VR1
REC. LEVEL CONTROL
VR101
AMP. PCB
J3, J103 PREAMP OUT JACK
J1, J101 AUX IN JACK
JACK PLATE
TO CHASSIS
MIC. JACK
J2
J102
REC. INDICATOR LAMP
LAMP 1
S7 MUTING SWITCH
H2
RECORD HEAD

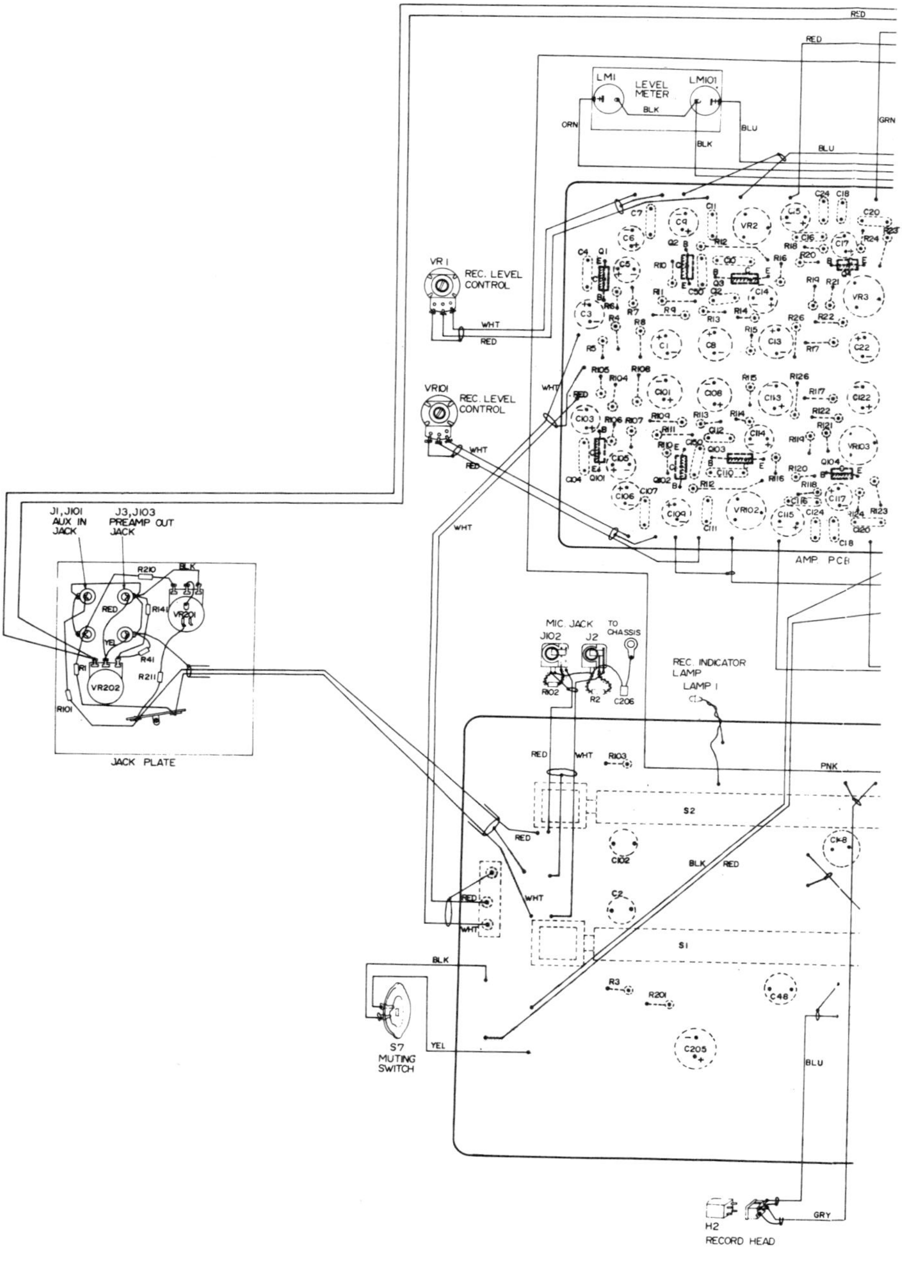
LEVEL METER
LM1
LM101
REC. LEVEL CONTROL
VR1
VR101
AMP. PCB
J1, J101 AUX IN JACK
J3, J103 PREAMP OUT JACK
JACK PLATE
MIC. JACK
J102
J2
TO CHASSIS
REC. INDICATOR LAMP
LAMP 1
S7 MUTING SWITCH
H2 RECORD HEAD
S1
S2

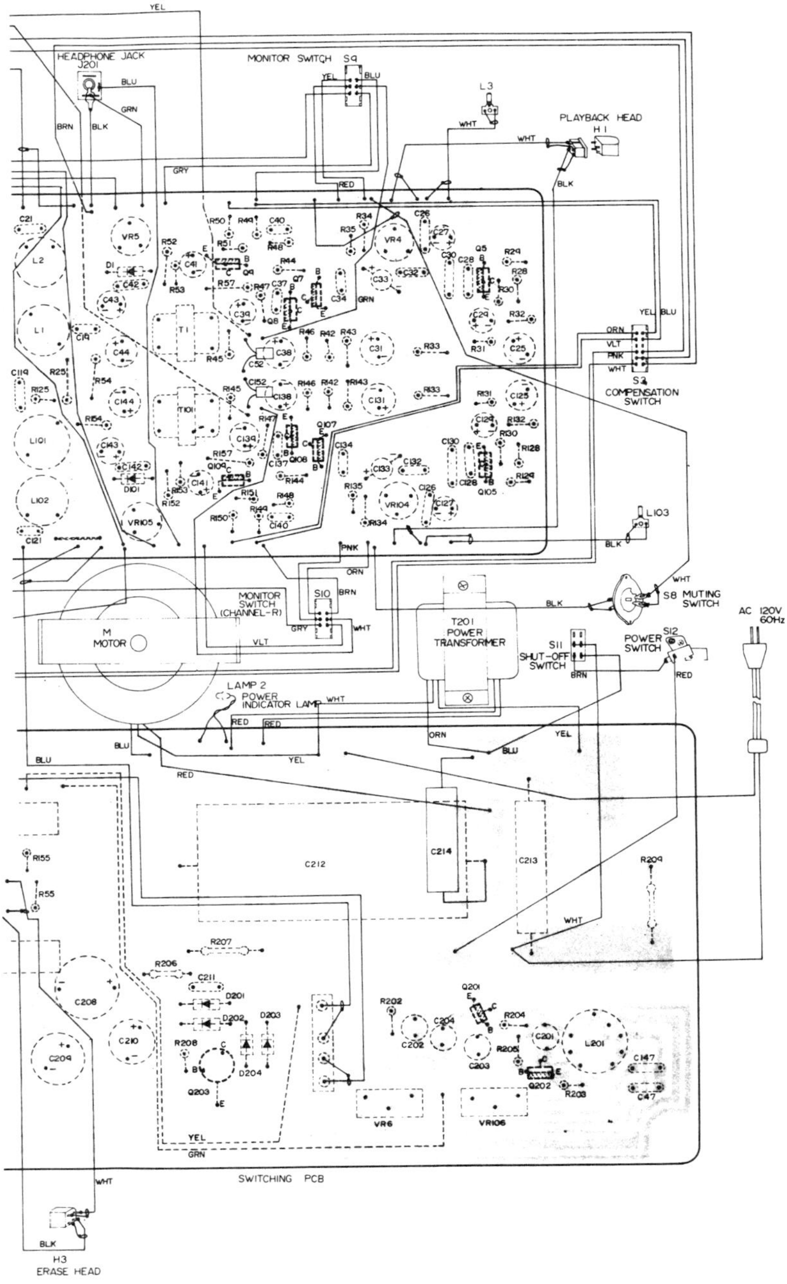
HEADPHONE JACK J201
MONITOR SWITCH S9
L3
PLAYBACK HEAD H1
S2 COMPENSATION SWITCH
L103
S8 MUTING SWITCH
AC 120V 60Hz
MONITOR SWITCH (CHANNEL-R) S10
M MOTOR
T201 POWER TRANSFORMER
S11 SHUT-OFF SWITCH
POWER SWITCH S12
LAMP 2 POWER INDICATOR LAMP
C212
C214
C213
SWITCHING PCB
H3 ERASE HEAD

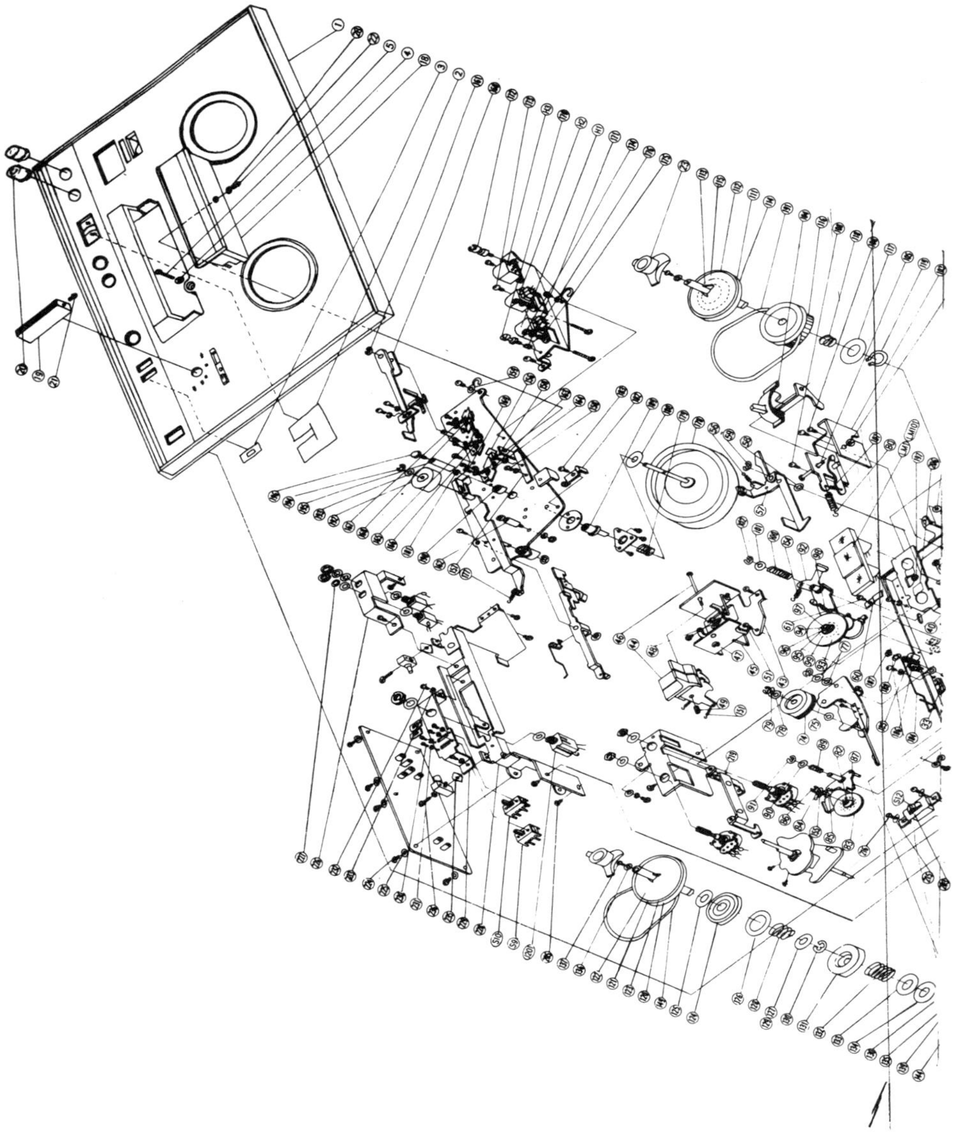

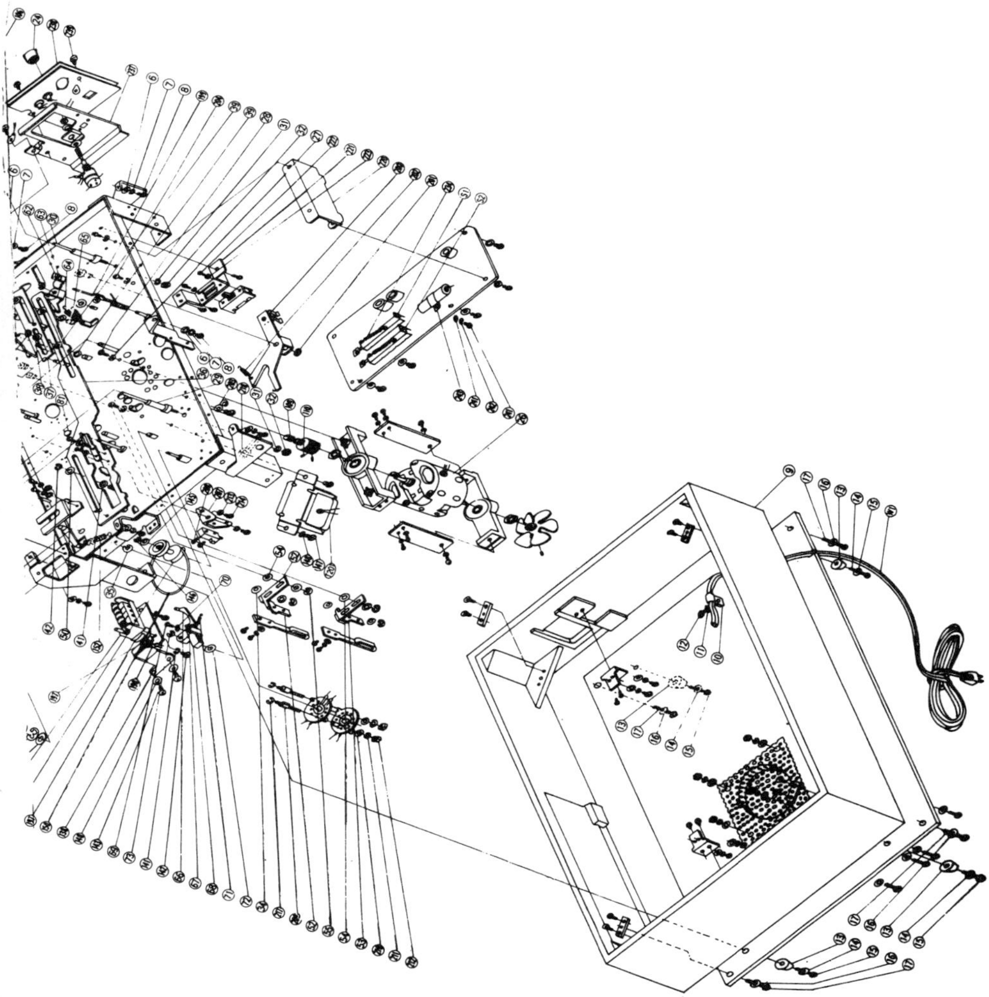

REPLACEMENT PARTS

Symbol No.	R.S. PART No.	MFR'S PART No.	Description		
CAPACITORS:					
C 1,101		0252623	Electrolytic	33μF	25V
C 2,102		0274113	Mylar	0.0022μF±20%	
C 3,103		0252621	Electrolytic	10μF	25V
C 4,104		0274014	Mylar	0.0033μF±10%	
C 5,105		0252223	Electrolytic	33μF	6.3V
C 6,106		0256004	Electrolytic	1.0μF	16V
C 7,107		0249515	Ceramic	510pF±5%	50V
C 8,108		0252631	Electrolytic	100μF	25V
C 9,109		0252615	Electrolytic	4.7μF	25V
C 10,110		0248718	Ceramic	56pF±10%	50V
C 11,111		0275115	Mylar	0.047μF±20%	
C 12,112		0275016	Mylar	0.068μF±10%	
C 13,113		0252223	Electrolytic	33μF	6.3V
C 14,114		0252613	Electrolytic	3.3μF	25V
C 15,115		0252613	Electrolytic	3.3μF	25V
C 16,116		0248736	Ceramic	330pF±10%	
C 17,117		0252615	Electrolytic	4.7μF	25V
C 18,118		0275012	Mylar	0.015μF±10%	
C 19,119		0275011	Mylar	0.01μF±10%	
C 20,120		0274015	Mylar	0.0047μF±10%	
C 21,121		0248722	Ceramic	82pF±10%	
C 22,122		0252631	Electrolytic	100μF	25V
C 24,124		0274015	Mylar	0.0047μF±10%	
C 25,125		0252525	Electrolytic	47μF	16V
C 26,126		0248736	Ceramic	330pF±10%	
C 27,127		0252621	Electrolytic	10μF	25V
C 28,128		0274111	Mylar	0.001μF±10%	
C 29,129		0252223	Electrolytic	33μF	6.3V
C 30,130		0276111	Mylar	0.1μF±20%	
C 31,131		0252625	Electrolytic	47μF	25V
C 32,132		0248734	Ceramic	270pF±10%	
C 33,133		0256004	Electrolytic	1.0μF	16V
C 34,134		0248736	Ceramic	330pF	
C 37,137		0248728	Ceramic	150pF±10%	50V
C 38,138		0252322	Electrolytic	22μF	10V
C 39,139		0252613	Electrolytic	3.3μF	6.3V
C 40,140		0275014	Mylar	0.033μF±10%	
C 41,141		0252615	Electrolytic	4.7μF	25V
C 42,142		0275111	Mylar	0.01μF±20%	
C 43,143		0252621	Electrolytic	10μF	6.3V
C 44,144		0252631	Electrolytic	100μF	25V
C 47,147		0247894	Ceramic	100pF±10%	50V
C 48,148		0221346	Styrol	3000pF	
C 50,150		0248728	Ceramic	150pF±10%	50V
C 52,152		0274112	Mylar	0.0015μF±20%	
C 201		0221337	Styrol	1200pF	
C 202		0221342	Styrol	2000pF	
C 203		0221342	Styrol	2000pF	
C 204		0221351	Styrol	4700pF	
C 205		0252621	Electrolytic	10μF	25V
C 206		0274111	Mylar	0.001μF±20%	
C 207		0275115	Mylar	0.047μF±20%	
C 208		0252631	Electrolytic	100μF	25V
C 209		0256008	Electrolytic	470μF	25V
C 210		0256008	Electrolytic	470μF	25V
C 211		0275111	Mylar	0.01μF±20%	
C 212		0256929	Metalized	1.2μF	
C 213		0215045	Paper	0.047μF	
C 214		0215045	Paper	0.047μF	
RESISTORS:					
R 1,101		0131765	Composition	390kΩ±10%	¼GF
R 2,102		0131671	Composition	1.5kΩ±5%	¼GF
R 3,103		0137851	Carbon film	1kΩ±10%	¼SD
R 4,104		0137956	Carbon film	270kΩ±10%	¼SD
R 5,105		0137909	Carbon film	47kΩ±10%	¼SD
R 6,106		0137905	Carbon film	22kΩ±10%	¼SD
R 7,107		0137862	Carbon film	8.2kΩ±10%	¼SD
R 8,108		0137803	Carbon film	150Ω±10%	¼SD
R 9,109		0137908	Carbon film	39kΩ±10%	¼SD
R 10,110		0137954	Carbon film	180kΩ±10%	¼SD
R 11,111		0137802	Carbon film	120Ω±10%	¼SD

Symbol No.	R.S. PART No.	MFR'S PART No.	Description		
R 12,112		0137955	Carbon film	220kΩ±10%	¼
R 13,113		0137910	Carbon film	56kΩ±10%	¼
R 14,114		0137908	Carbon film	39kΩ±10%	¼
R 15,115		0137860	Carbon film	5.6kΩ±10%	¼
R 16,116		0137853	Carbon film	1.5kΩ±10%	¼
R 17,117		0137811	Carbon film	680Ω±10%	¼
R 18,118		0137907	Carbon film	33kΩ±10%	¼
R 19,119		0137912	Carbon film	82kΩ±10%	¼
R 20,120		0137903	Carbon film	15kΩ±10%	¼
R 21,121		0137861	Carbon film	6.8kΩ±10%	¼
R 22,122		0137854	Carbon film	1.8kΩ±10%	¼
R 23,123		0137862	Carbon film	8.2kΩ±10%	¼
R 24,124		0137905	Carbon film	22kΩ±10%	¼
R 25,125		0137769	Carbon film	68Ω±10%	¼
R 26,126		0137808	Carbon film	390Ω±10%	¼
R 28,128		0137956	Carbon film	270kΩ±10%	¼
R 29,129		0137909	Carbon film	47kΩ±10%	¼
R 30,130		0137905	Carbon film	22kΩ±10%	¼
R 31,131		0137862	Carbon film	8.2kΩ±10%	¼
R 32,132		0137805	Carbon film	220Ω±10%	¼
R 33,133		0137905	Carbon film	22kΩ±10%	¼
R 34,134		0137851	Carbon film	1.0kΩ±10%	¼
R 35,135		0137909	Carbon film	47kΩ±10%	¼
R 41,141		0131659	Composition	680Ω±10%	¼
R 42,142		0137954	Carbon film	180kΩ±10%	¼
R 43,143		0137802	Carbon film	120Ω±10%	¼
R 44,144		0137955	Carbon film	220kΩ±10%	¼
R 45,145		0137811	Carbon film	680Ω±10%	¼
R 46,146		0137860	Carbon film	5.6kΩ±10%	¼
R 47,147		0137853	Carbon film	1.5kΩ±10%	¼
R 48,148		0137911	Carbon film	68kΩ±10%	¼
R 49,149		0137857	Carbon film	3.3kΩ±10%	¼
R 50,150		0137856	Carbon film	2.7kΩ±10%	¼
R 51,151		0137905	Carbon film	22kΩ±10%	¼S
R 52,152		0137810	Carbon film	560Ω±10%	¼S
R 53,153		0137808	Carbon film	390Ω±10%	¼S
R 54,154		0137759	Carbon film	10Ω±10%	¼S
R 55,155		0134379	Composition	3.3kΩ±10%	½
R 57,157		0137808	Carbon film	390Ω±10%	¼S
R 201		0137759	Carbon film	10Ω±10%	¼S
R 202		0137768	Carbon film	56Ω±10%	¼S
R 203		0137768	Carbon film	56Ω±10%	¼S
R 204		0137905	Carbon film	22kΩ±10%	¼S
R 205		0137905	Carbon film	22kΩ±10%	¼S
R 206		0134300	Composition	82Ω±10%	½C
R 207		0134297	Composition	47Ω±10%	½C
R 208		0137810	Carbon film	560Ω±10%	¼S
R 209		0134413	Composition	2.2MΩ±10%	½G
R 210		0137909	Carbon film	47kΩ±5%	¼G
R 211		0131770	Composition	680kΩ±5%	¼G
VR1,101	P-0660	0153152	Variable	10kΩ (REC.LEVEL)	
VR2,102	P-0657	0151888	Semi variable	47kΩ (REC.OUTPUT LEV	
VR3,103	P-0658	0151887	Semi variable	22kΩ (REC.CURREN	
VR4,104	P-0657	0151888	Semi variable	47kΩ (P.B.OUTPUT	
VR5,105	P-0652	0151886	Semi variable	10kΩ (LEVEL METE	
VR6,106	P-0659	0151829	Semi variable	330kΩ (BIAS CURREN	
VR 201		0151756	Variable	10kΩ (S-O-S, VOLUM	
VR 202		0151370	Variable	10kΩ (PREAMP OU	
TRANSISTORS:					
Q 1,101		5320024	Transistor	2SC458D(LG)	
Q 2,102		5320064	Transistor	2SC458D	
Q 3,103		5320064	Transistor	2SC458D	
Q 4,104		5320064	Transistor	2SC458D	
Q 5,105		5320024	Transistor	2SC458D(LG)	
Q 7,107		5320024	Transistor	2SC458D(LG)	
Q 8,108		5320064	Transistor	2SC458D	
Q 9,109		5320613	Transistor	2SC1213C	
Q 201		5320613	Transistor	2SC1213C	
Q 202		5320613	Transistor	2SC1213C	
Q 203		0573024	Transistor	2SB370A	
D 1,101		0575001	Diode	1N34A	

Symbol No.	R.S. PART No.	MFR'S PART No.	Description	
D 201		5330104	Diode	VO6A
D 202		5330104	Diode	VO6A
D 203		5330104	Diode	VO6A
D 204		5330104	Diode	VO6A
			TRANSFORMERS:	
1,101	TA-0030	5250671	Headphone	
201	TA-0257	5210603	Power transformer	
			COILS:	
1,101	CA-2543	5270031	Variable	6mH
2,102	CA-2544	5270001	Variable	33mH
3,103		5270061	Variable	
201	CA-4254	5260022	Oscillator	
			for Final assembly	
1	Z-0363	6124313	Decoration plate	
2		7732161	Record button sheet	
3		7779541	Switch cover	
4		0941277	Decoration plate washer	
5		7771274	Decoration plate screw	
6		8811114	Washer-3mmø	
7		8813124	Spring washer-3mmø	
8		8711408	Pan head screw-3mmø×8mm	
			(7 req'd) for escutcheon mounting	
9	Z-0364	6117806	Cabinet assembly	
10		0948483	Staple	
11		8812114	Washer-3mmø	
12	HS-0186	8751413	Wood screw-3.1mmø×13mm	
			for power cord mounting	
13	F-0073	7662191	Rubber leg	
14		8812116	Washer-4mmø	
			(2 req'd) for rubber leg mounting	
15		8711625	Pan head screw-4mmø×25mm(4 req'd)	
			for rubber leg, cabinet mounting	
16		0645201	Special washer-4mmø	
17		8711618	Pan head screw-4mmø×18mm(4 req'd)	
			cabinet mounting	
18	RA-8006	6164944	Head cover assembly	
19	K-0869	6263532	Selector lever	
		7778701	Rubber washer	
		6263532	Knob assembly	
20	K-0870	6263523	Record level knob	
21		7777141	Set screw	
22		0747306	Binding screw-2.6mmø×6mm	
23		8813123	Spring washer-2.6mmø	
			for head cover mounting	
24	K-0871	6701842	Volume knob assembly	
25	RA-1113	7666861	Reel cap assembly	
		8711404	Pan head screw-3mmø×4mm	
			for Chassis assembly	
26	M-4168	5570034	Motor assembly	
27		8813126	Spring washer-4mmø	
28		8711610	Pan head screw-4mmø×10mm	
29	RA-4848	7500502	Take up reel shaft	
30	RA-4849	7500522	Rewind reel shaft	
31		8821116	Nut-4mmø	
32		8813126	Spring washer-4mmø	
33		7732791	Oil sheet	
34	RA-4850	7212022	Function cam assembly	
35		7230903	"E" ring-4mmø	
36	RA-4851	7175511	Function bar assembly	
37	RA-4852	7178721	Tape speed chage-over arm assembly	
38		0948603	Fiber washer	
39		0918544	Fiber washer	
40		7230903	"E" ring-4mmø	
41		7711403	Fiber washer-8.1mmø×12mmø	
			for function bar mounting	
42		7230904	"E" ring	
43	H-0882	7178751	Record lever holder	
44	K-0872	6263413	Record button	
45	RA-4853	7505954	Record lock rod	
		7779521	Shield plate (J)	
46	RA-4854	7505961	Record lever shaft	
47		7230903	"E" ring-4mmø	
48		6311931	Timing plate spring	
			for record lock	
49		6319431	Spring for shaft	
50		0948810	Fiber washer-8.1mmø×12mmø	

Symbol No.	R.S. PART No.	MFR'S PART No.	Description
51		8781438	Tapping screw-3mmø×8mm
52	RA-4855	7178778	Record slide plate (L) assembly
53	RA-4856	7178793	Record slide plate (R) assembly
54		0948544	Fiber washer
55		7230903	"E" ring-4mmø
56	RA-5319	6319011	Slide plate spring
			for P arm function plate
57	RA-4857	7178821	Pressure function plate arm
58		0948544	Fiber washer
59		7230903	"E" ring-4mmø
60		7180111	Pressure function plate holder
61		8781438	Tapping screw-3mmø×8mm
62	RA-4862	7216991	Brake lever (L) assembly
63		0971048	Brake rubber
64	RA-5320	0948202	Rewind brake spring
65	RA-5321	0662131	Push lever spring for brake lever
66	RA-4863	7211491	Brake lever (R) assembly
67		0971048	Brake rubber
68		7663213	Rubber cushion
69		8781440	Tapping screw-3mmø×10mm
			for rubber cushion mounting
70		8811114	Washer-3mmø
			for rubber cushion mounting
71		6316851	Brake spring (R)
		7779531	Shield plate (P)
72		0948544	Fiber washer
73		7230903	"E" ring-4mmø
74	RA-2210	6721582	Rewinding idler
75	RA-4858	7178872	Rewinding idler assembly
76		0948595	Nylon washer
77		0948544	Fiber washer
78		0948601	Idler oil washer
79		0948633	Rewinding idler washer
80	RA-5322	6313121	Rewinding idler lever spring
81		0630564	Flywheel pad
82	RA-4859	7178901	Function idler lever assembly
83	RA-2211	0971150	Idler wheel
84		0948595	Nylon washer
85		0948601	Idler oil washer
86		7230903	"E" ring-4mmø
87	RA-1114	0971105	Idler cap
88		0958067	Felt for push button
89		0948631	Function idler change-over spring
90		0948675	Fiber washer
91		7230903	"E" ring-4mmø
92	RA-4864	7178941	Pressure idler lever assembly
93	RA-2212	6380251	Pressure idler
94		0948727	Pressure nylon washer
95		0948601	Idler oil washer
96		7230902	"E" ring-3mmø
97	RA-1115	0971215	Idler cap
98	RA-5323	0662128	Pressure idler spring
99		0958067	Felt for push button
100	RA-5324	6314271	Selector spring
101		0948675	Fiber washer
102		7230903	"E" ring-4mmø
103		7178982	Tape speed chage-over holder assembly
104		6263421	Tape speed change-over button
105		7506041	Tape speed change-over shaft
106		7230903	"E" ring-4mmø
107	RA-5325	0662126	Function idler lever spring
108		8781438	Tapping screw-3mmø×8mm
			(3 req'd) for tape speed change-over
			holder mounting
109	RA-2213	6342341	Motor pulley
110		7771252	Set screw
111	RA-1116	6410602	Rewind reel stand
112	H-0883	7662111	Reel stand rubber
113		6604851	Reel spool
114	H-0884	7712371	Reel stand sheet
115	RA 6064	0948665	Take up pulley felt
116	RA-2214	6712155	Rewind pulley
117		7779111	Washer
118	RA-5326	6316711	Reel stand spring
119		0948662	Snap
120	RA-1117	6410633	Take up reel stand
121		7662111	Reel stand rubber
122		6604851	Reel spool
123		7712371	Reel stand sheet
124	RA-2215	6700741	Take up pulley (A)

Symbol No.	R.S. PART No.	MFR'S PART No.	Description
125	RA-6065	0948806	Take up pulley felt
126	HW-0140	7779131	Take up slip washer
127	HW-0141	7779141	Take up washer
128	RA-5327	6310301	Reel stand sping
129		7710155	Cupring washer
130		7230908	"E" ring
131	RA-2216	6700752	Take up pulley (B)
132	RA-5328	6313922	Take up spring
133	H-0885	6719601	Take up collar
134		7660241	Nylon washer
135		0948595	Nylon washer
136		0948482	Fiber washer
137		7771312	Special screw
138	H-0886	7218211	Counter holder
139	D-2038	5550243	Counter M480
140		8811114	Washer-3mm ø
141		8813124	Spring washer-3mm ø
142		8711406	Pan head screw-3mm ø × 6mm
143		8781436	Tapping screw-3mm ø × 6mm (2 req'd) for counter holder mounting
144	RA-2217	6340341	Counter pulley
145		0636553	Rewind washer
146		7230903	"E" ring-4mm ø
147		8711606	Pan head screw-4mm ø × 6mm
148		8813126	Spring washer-4mm ø (2 req') for trans. holder mounting
149	B-6165	6971193	Counter belt
150	B-6166	0971226	Counter belt
151		6319261	Pad spring
152	RA-5329	6319121	Brake lever spring for adjusting plate
153	RA-5330	6312081	Pressure spring
154	RA-5331	6312101	Pressure lever spring
155		0662193	Pause lever spring for function idler lever
147		8711606	Pan head screw-4mm ø × 6mm
		8811116	Washer-4mm ø
148		8813126	Spring washer-4mm ø
156		7714811	Holder sheet
157		7195221	Cover holder
158		8711406	Pan head screw-3mm ø × 6mm
159		0923972	Staple
160	RA-4860	7179112	T arm assembley
161		7230903	"E" ring-4mm ø
162	RA-4865	7586332	Tape guide
163		7179152	Pad plate assembly
164		7715302	Pad
165		7191311	Pad plate assembly
166		0948835	Pad
167		7191341	Pad plate assembly
168		0948756	Pad
169		7771873	Screw with spring washer-3mm ø (6 req'd) for pad plate mounting
H1, H2	H-4136	5440531	Record playback head
H3	H-4137	5440512	Erase head
170	RA-4867	0944741	Tape guide (L)
171	RA-5332	0948154	Tape guide spring
172	RA-4866	7505214	Tape guide (L)
173		0711306	Pan head screw-2.6mm ø × 6mm
174		0821233	Nut-2.6mm
175	RA-5333	6318371	Head adjust spring
176		8811233	Washer-2.6mm ø
177	RA-5334	6312081	Pressure spring for pressure roller
178	RA-3090	6370331	Flywheel assembly
179	RA-5335	6313332	Take up spring
180		7715551	Function washer
181		7230903	"E" ring-4mm ø
182	H-0901	7180121	Function idler stopper
183		8711408	Pan head screw-3mm ø × 8mm (4 req'd) for sub chassis mounting
184		6313783	Shaft of pin
185	H-0887	7506131	Balancer
186		7732801	Oil washer
187		7230902	"E" ring-3mm ø
188		0637119	Fiber washer
189	HW-0142	0948634	Flywheel washer
190	RA-1118	0941183	Oil cap
191	B-6167	0971126	Rewind belt
192	RA-2218	6380171	Pressure roller
193	HW-0143	0636553	Rewind washer
194		7230902	"E" ring-3mm ø
195	RA-7110	6360028	Capstan
S 1, S 2	S-2060	5620602	RECORD switch(Slide)
		7213241	Heat sink plate for Q203
	J-6100	5660411	4P miniature plug
	J-6101	5660401	3P miniature plug
	J-0317	5680441	1 pole terminal pin
	J-6102	0542109	Plug-2.35mm ø
		8711408	Pan head screw-3mm ø × 8mm

Symbol No.	R.S. PART No.	MFR'S PART No.	Description
196	HS-0187	6360011	Capstan screw
197	H-0888	0971002	Lamp cushion
198	L-0278	0594110	Pilot lamp
199	RA-5336	6313052	Balance spring
200	RA-4861	7179221	Equalizer arm assembly
201		7230903	"E" ring-4mm ø for equalizer arm mounting
202	RA-5337	0639391	Motor spring for equalizer arm
		7714951	Switch sheet
203		0923972	Staple (2 req'd) for power trans.
204		8711406	Pan head screw-3mm ø × 6mm (4 req'd) for P.C.B. holder mountin
205		8781438	Tapping screw-3mm ø × 8mm (3 rec for equalizer holder, jack holder, (2 req'd) for P.C.B. holder, mic ja holder, counter holder mounting
LM1 LM101	M-0140	5550253	LEVEL meters
206	H-0889	7666011	Level meter cushion
207	H-0890	7183263	Switch holder
208		0745314	Binding screw-2.6mm ø × 14mm
	S-6008	0539135	Micro switch
209	H-0891	7585853	Switch shaft
210		8811113	Washer-2.6mm ø
211		8813123	Spring washer-2.6mm ø
212		0821233	Nut-2.6mm ø (2 req'd) for rotary switch mountin
213		8813124	Spring washer-3mm ø
214		8711406	Pan head screw-3mm ø × 6mm (2 req'd) for switch holder mountin
215		8811114	Washer-3mm ø
216		7771873	Screw with spring washer-3mm ø (2 req'd) for power switch mountin
S 12	S-3009	5630363	POWER switch, rocker
217		8813124	Spring washer-3mm ø for switch shaft mounting
	S-1034	5610251	Rotary switch
	S-1035	5610412	Rotary switch
218	H-0892	0043793	Bushing
W1	W-1293	5741666	Power cord
		7114561	Back tension assembly
			for Printed Circuit Board assembly
		5742078	Cord for between P.C.B. jack J3
		5742076	Cord for between P.C.B. jack earth
		5742077	Cord for between P.C.B. jack J103
		5742096	Cord for between selector P.C.B. S7
		5742095	Cord for between selector P.C.B. P.C.B. (+) (−)
	J-6098	5650512	4P miniature socket
	J-6099	5651011	3P miniature socket
		6712681	Fuse band
219	H-0893	7180523	P.C.B. holder assembly
220		7180161	P.C.B. pad
221	H-0894	7172001	Equalizer holder
	S-2059	5620912	Slide switch
222		0711304	Pan head screw-2.6mm ø × 4mm
223		8813123	Spring washer-2.6mm ø
224		8745406	Binding screw-3mm ø × 6mm
225		7711854	Fiber washer (4 req'd) for P.C.B. mounting
226	H-0895	7241701	Microphone jack holder
227		7778352	Fiber washer
	J-0314	5671142	MIC jack
228		7229154	P.C.B. holder assembly
229	H-0896	7241601	Jack holder
J 201	J-0319	5670282	Headphone jack
S 9, S 10	S-2060	5620183	Monitor switch(Slide)
230		7771854	Screw with spring washer-2.6mm ø
231		8811113	Washer-2.6mm ø
232		7771874	Screw with spring washer-3mm ø
233		8811114	Washer-3mm ø
234		0662266	Function bar spring for slide switch
	H-0897	0544186	Lug terminal
	H-0899	5650261	Head socket
	CA-2545	5270061	Balance coil-hum
235		7663721	Rubber sheet
236		8715316	Pan head screw-2.6mm ø × 16mm (2 req'd) for balance coil mounting
237	H-0900	7191521	Jack holder
238	H-0898	6215921	Jack sheet
239		8745406	Binding screw-3mm ø × 6mm
241		8811114	Washer-3mm ø
242		8813124	Spring washer-3mm ø
243		8821114	Nut-3mm ø
		5671301	Jack plate assembly
		8812116	Washer-4mm ø (2 req'd) for rubber base mounting
		7735991	Insulating plate

Chapter 5

Tape Cassette Transport Mechanisms: Service and Repair

Since cassette players are usually portable and thus subject to much use (and rough treatment), you will get plenty of requests to repair them.

Figure 5-1. Cassettes are easily inserted into a machine. (Courtesy of 3M)

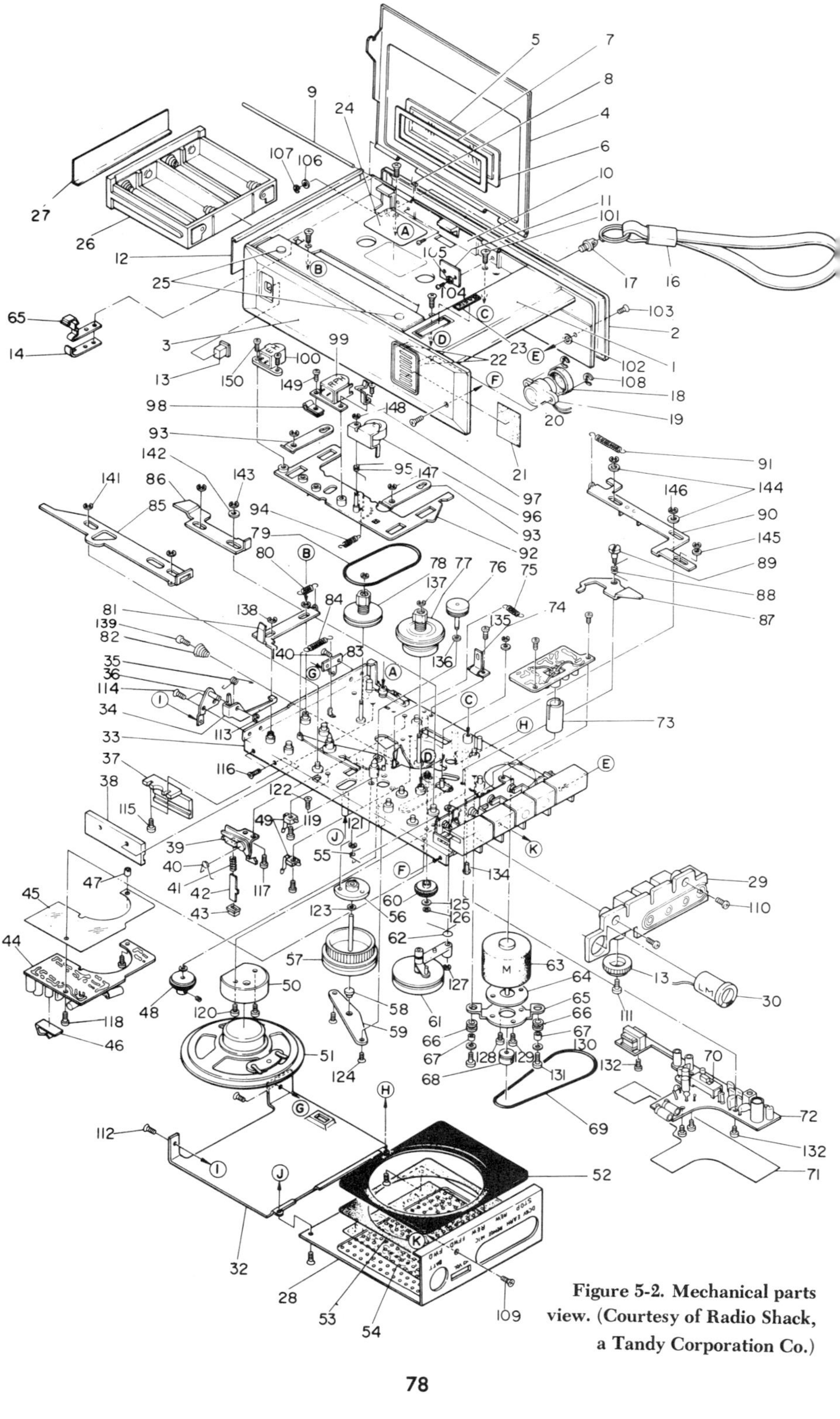

Figure 5-2. Mechanical parts view. (Courtesy of Radio Shack, a Tandy Corporation Co.)

Principles of cassette equipment (moving tape past heads)

Cassette recorder/players generally use only one motor which drives both the tape reels in the cassette and the capstan. The tape is moved from one reel to the other in the cassette; then the casette may be turned over and the process repeated. Recorded material thus appears on one half of the tape in each direction. Tape travel is maintained at a steady speed by the capstan pressing it (during movement) against a rubber wheel. Belts and drive wheels are used to rewind the tape, if desired. Tape travels past the heads at 1-7/8 ips only; no other speeds are used.

Basic operation modes are *play* and *record* (although some models, particularly those planned for use in cars, feature *play* only), *fast-forward*, and *rewind*. There are also switches to set up different electrical circuits (stereo or mono, CrO_2 tape or normal tape) and a device to eject the cassette after play.

Cassette player/recorders operate in the same basic manner as do open-reel machines. In the record, playback, or erase modes, the tape is placed against the head, the pinch roller is put in contact with the capstan, and current is applied to the motor(s). In fast-forward and rewind modes, the tape is released from the heads, the capstan is freed and current is supplied to the motor. Belts and drive wheels are used to change modes in most recorders; some make use of additional motors, in addition to the belts and pucks.

You'll find that most of your servicing time will be taken up with getting at the defective parts because the equipment is generally compact. Some parts will require periodic replacing; others, like the motor, will seldom burn out.

Motors in battery sets are, of course, dc motors, in which the speed is changed by voltage changes. Good quality players designed for use on household current will most often have hysteresis-synchronous motors for precise tape speed under varying conditions (in such motors the speed is controlled by current frequency). Induction motors (in which voltage changes control the speed) are also used. Most electronics technicians look for the source of malfunctions in other components and suspect the motors last.

The *capstan* (which pulls the tape past the heads at a steady speed) is the source of much trouble in cassette players primarily

Figure 5-3. The drive mechanism of one popular tape machine.

because the shaft on which it is mounted is rather easily bent. In making replacements, you have to be careful that you don't damage the shaft during installation and end up having to do the whole job over again.

Belt and *rubber wheel drives* (which connect fast-forward, rewind and other modes) are a frequent source of trouble. Keep them clean (use denatured alcohol) and replace those that are damaged (belts which are flaking or wheels with flat edges).

Brakes (usually a pad rubbing a metal wheel) often need adjustment. The *take-up reel* rotates at varying speeds to wind the tape evenly (slower as more tape is wound on the reel); this variation is controlled by a *clutch* which sometimes requires adjustment or replacement.

The *cassette* itself will often be the source of problems. Wheel hubs which do not function correctly can cause the tape to bind, travel unevenly, or not move at all. In addition, worn or damaged pressure pads in the cassette itself can cause wow-and-flutter and other distortion. Also, if the cassette's record interlock tabs are

Figure 5-4. Watch for tell-tale signs of belt wear like these rubber flakes.

Figure 5-5. Check rubber wheels for flatness.

removed (two pieces of plastic on the edge opposite from the one containing the pressure pads), the recorder will not record. This is a safety device to make it impossible to erase recordings unless you choose to do so. However, customers unaware of this feature will often complain that their recorders have broken down.

How to disassemble any machine in short order

The quickest way to take a machine apart is to check the service manual for that specific recorder. If one is not at hand, it is best to think before unscrewing everything in sight. Otherwise, you are wasting your valuable time.

Usually, you will have to remove both the bottom *and* the top of the recorder/player to get at essential components. Start with the bottom. As rule of thumb, you'll find that the most obvious screws are not the ones you want—generally, the three to six screws holding the base to the mechanism are at or near the corners, but are sometimes hidden under rubber bumpers or a plate or shield (unskilled hands will be less likely to tamper with them and there your job will be easier).

Remove one screw and look to see if it is the type of screw and the length you would expect before proceeding to remove them all. (If you remove something you shouldn't have, replace it before proceeding with the repairs. Otherwise you'll forget to do so later and waste more time.) Put all the screws in a cup; you won't lose them that way.

To remove the top, take the knobs off (either set-screw or pull-off knobs are obviously simple to remove). Unscrew three or four screws and pull off the cover.

Put the chassis down on a soft cloth and you're ready to find the source of the trouble.

A useful checklist to pinpoint problems in play/record, fast-forward, rewind

Test the equipment yourself after noting the customer's complaint. Use a quality cassette and a dependable electric cord or fresh batteries to make certain that malfunctions are in the equipment itself. Put the machine into various modes. After ascertaining the problem, use the following checklist to narrow the causes to

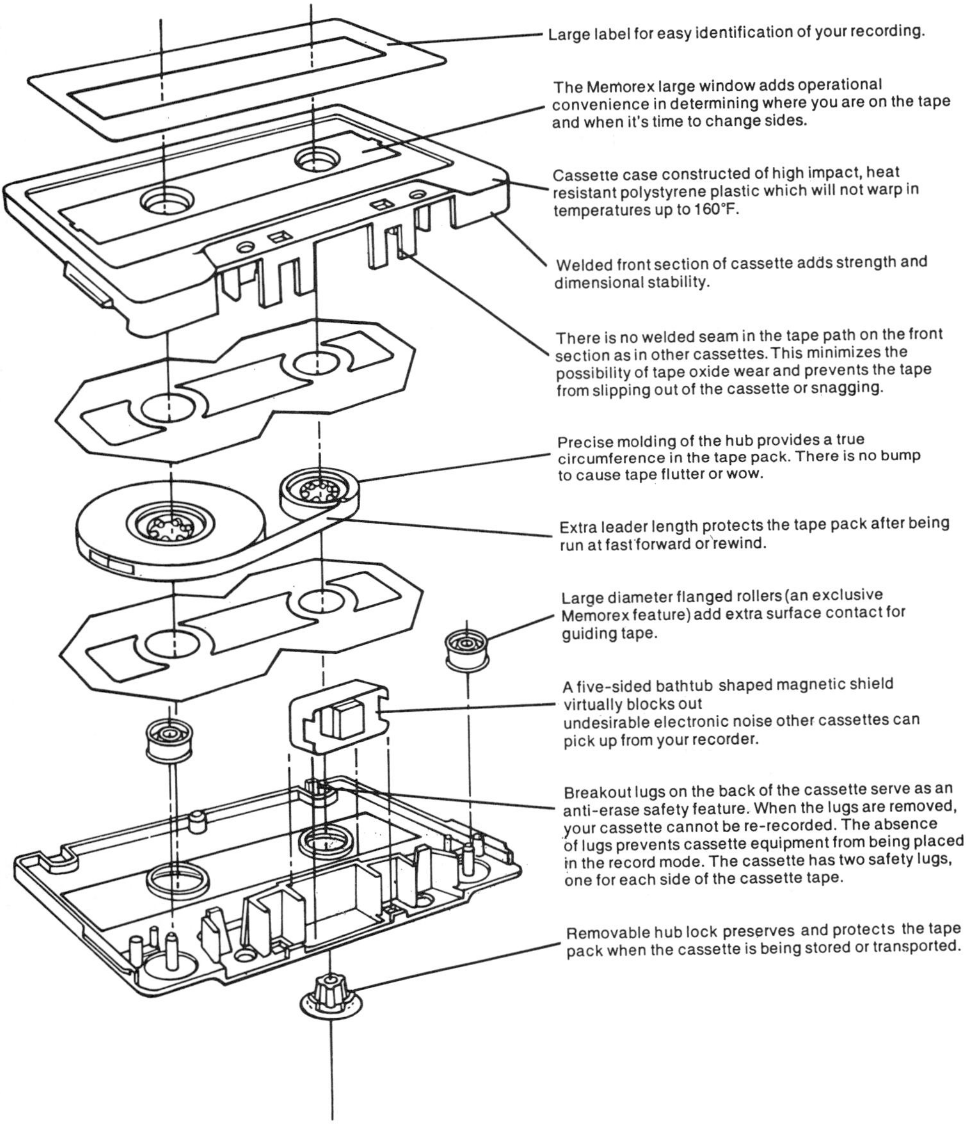

Figure 5-6. What a cassette looks like. (Courtesy of Memorex)

those you can correct quickly. Repair the trouble by doing *one* of the following:

1. *If the motor doesn't run*
   a. check to see that the external microphone switch is on.
   b. check and replace the on-off switch (if defective).

Figure 5-7. Motors seldom need replacement.

c. replace the capacitor at the motor.
d. replace the motor.

2. *If the capstan doesn't turn*
   a. replace the broken drive belt.
   b. tighten the loose screw on the pulley (if necessary).
   c. replace the motor.

3. *If the tape starts too slowly*
   a. use a different (and dependable) cassette.
   b. replace the defective drive belt.
   c. straighten (or replace) the bent drive shaft.

4. *If the speed is too fast*
   a. replace the defective motor governor.

5. *If the tape speed is too slow*
   a. use a different (and dependable) cassette.
   b. stop belt from slipping.
   c. stop the take-up reel from slipping.
   d. stop the pressure roller from slipping.
   e. replace the defective drive wheel.
   f. replace the defective flywheel.
   g. repair or replace the defective capstan.
   h. replace the defective thrust cap of the capstan housing.
   i. replace the malfunctioning motor.

6. *If the tape breaks or jams*
   a. use a dependable quality cassette.
   b. repair or replace the defective capstan housing.
   c. correct or replace the bent or misshaped roller plate.
   d. realign the heads (see Chapters 7, 8 and 9).

7. *If the tape spills*
   a. replace the broken drive belt.
   b. clean the oily or dirty pinch roller.
   c. adjust the pinch roller pressure according to specifications.
   d. adjust the take-up torque according to specifications.

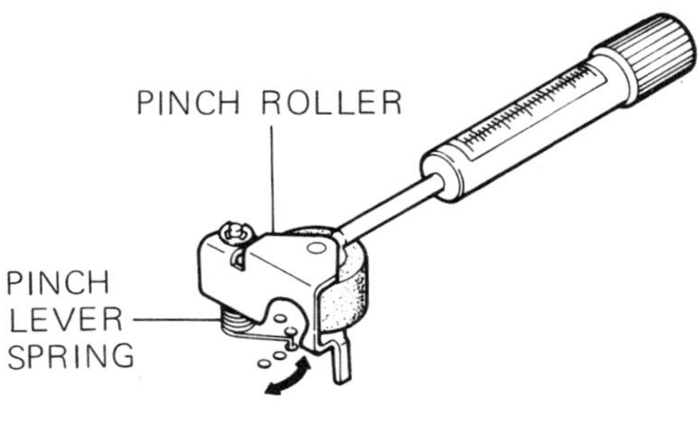

Figure 5-8. How to measure the pressure roller (pinch roller) pressure force. (Courtesy of Radio Shack, a Tandy Corporation Co.)

8. *If the spindles do not revolve in take-up mode*
   a. replace the defective cassette.
   b. stop the pressure roller from slipping.
   c. stop the drive belt from slipping.
   d. adjust the brake (1/64 inch clearance is common).
   e. adjust the rewind torque (which is too low) to meet specifications (usually 30 to 60 grams).

9. *If spindles do not move in fast-forward or rewind mode*
   a. replace the defective cassette.
   b. stop the drive belt from slipping.
   c. replace the broken belt.
   d. repair or replace the inoperative leaf switch.
   e. adjust the inadequate tension on the fast-forward or rewind spring according to specifications (usually about 40 to 50 grams).
   f. adjust the brake so that it doesn't touch the turntable (approximately 1/64 inch is adequate clearance).

10. *If wow-and-flutter occurs*
    a. replace the defective cassette.
    b. replace defective belts or drive wheels.
    c. adjust the take-up torque to proper specifications (usually 30 to 60 grams).
    d. correct the thrust gap to original specifications.
    e. align the heads (see Chapters 7, 8 and 9).
    f. lubricate flywheel bearing sparingly.
    g. adjust pinch roller spring to specifications (usually about 40 grams).
    h. replace the defective motor.
    (Since wow-and-flutter is often hard to pin down, you might need to perform two or more of these operations.)

11. *If the automatic stop mechanism doesn't work*
    a. replace the broken or inoperative lock spring.
    b. clean or replace the pinch roller.
    c. replace the defective solenoid.

In general, clean all drive surfaces to eliminate oil, dust and dirt. Use quality cassettes and make sure that the cassette you insert in the machine for a test is wound correctly (turn the hub tight with a pencil, but don't force it). Use accessories (microphones, for example) which you know to be good.

You may find it necessary to take readings of the torque or mechanical drive force with a gauge. (Some electronics technicians develop a feel for torque that is uncanny by stalling the spindles in various modes with their fingers. Try it.) Obviously, it's best to rely on specifications which are given in service manuals, but usually you'll find the following torques to be about average for cassette machines: *rewind torque*—40 grams, *fast-forward torque*—40 grams, *take-up torque*—30 to 60 grams, *capstan pull force*—40 to 60 grams. These measurements are made by attaching the gauge to the shaft of the various drive mechanisms. Adjust tension by moving the spring to another hole.

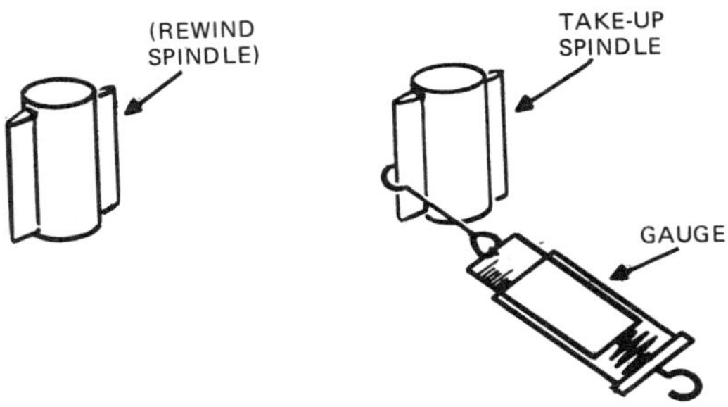

Figure 5-9. How to measure fast forward, rewind, and take-up tension. (Measure while placed in the operating mode.)

Be careful not to overlubricate. Most manufacturers claim that mechanical parts are factory coated with grease and should not require further lubrication. If you find it necessary to use grease, be careful that you do not get it on any of the friction drive parts or on the heads. (If, in spite of everything, you do, clean it off with denatured alcohol.)

After adjustments of screws and similar parts, apply a small amount of locking compound (or a dab of nail polish) to the adjusted parts so that vibration won't knock them out of adjustment.

Finally, do not use a magnetized screwdriver to make any adjustments.

Common sense ways to avoid trouble with cassette mechanisms

If you keep three common sense rules in mind, you'll find dollars coming your way from satisfied customers who know you fix things right—the first time.

Common Sense Rule #1: Keep it clean! A few minutes spent cleaning the heads of each piece of equipment to reach your bench will pay off. Use a Q-tip soaked in denatured alcohol or a commercially available cleaning cassette. (In addition, spend about one minute demagnetizing the heads as well). Spend a couple more minutes wiping off the outside case with a cloth and spray cleaner. Your customers will be pleased to get a recorder back that not only sounds fine but looks like new! Interior cleaning may also save you from having to do the same job over (without getting paid) because that newly installed belt keeps slipping or catching on an unclean reel!

Common Sense Rule #2: Be gentle! Take care in moving the player on your workbench; set it down on a soft cloth. Be careful not to force anything. For example, with some players (particularly ones which operate vertically), you might insert the cassette incorrectly and find that the tape goes behind the capstan (instead of between the capstan and the pinch roller). You'll either wind up with no sound or else you'll be unwinding tape from a capstan that looks like a tangled fishing reel. You might find that the cassette won't snap into the compartment and thus won't play (because of rough handling). You might even find that the tape pulls out of the cassette if you yank it from the machine (often due to a damaged slot) and you'll wind up with another mess. Rough handling is for owners only—not for servicemen.

Common Sense Rule #3: Treat it to the best! From selecting trouble-free quality tape cassettes to replacing components, choose only reliable brand-name merchandise. The pennies you save using bargain parts will be lost in the dollars you lose in the aggravation of doing the job over. Go first class all the way!

GENERAL ELECTRIC

SERVICE INFORMATION

MODEL
M8445A
PORTABLE
TAPE
RECORDER

FILE - M

CAUTION: THIS MANUAL IS DESIGNED FOR USE BY QUALIFIED ELECTRONIC TECHNICIANS ONLY. CONSUMER USERS ARE URGED TO CONTACT QUALIFIED FACTORY AUTHORIZED SERVICE FACILITIES FOR REPAIRS.

GENERAL SPECIFICATIONS

GENERAL SPECIFICATIONS
POWER OPTIONS **AC –** 120vAC Line Cord **DC –** (5) 1.5vDC "C" Cell Batteries **DC –** RT9508 Auto Lighter Adaptor
TAPE SPEED – 1-7/8 i.p.s. ±3% from 5.5 to 7.5vDC
IMPEDANCES **INPUT –** 3.3K **OUTPUT –** 8 ohms **EXT. MIKE –** 600 ohms **EARPHONE –** 8 ohms **SPEAKER –** 8 ohms **PATCHCORD MATCHING –** 200 to 600 ohms
FREQUENCY RESPONSE **PRE-RECORDED CASSETTE –** 150 to 6KHz ±6db **MIKE RECORDED CASSETTE –** 200 to 4KHz ±5db
POWER OUTPUT – 500 milliwatts min.
RECORD BIAS – DC
ERASE VOLTAGE – DC
USER MANUAL – 247A3305 Rev. 0

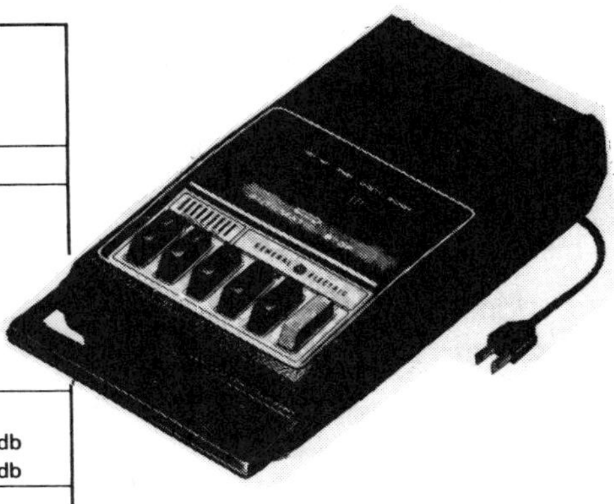

GENERAL INFORMATION

General Electric Model M8445A is a portable cassette Tape Recorder that is mechanically similar to Model M8455A and electrically similar to Model M8455A. It operates on 7.5vDC which can be supplied by– 1.) 5 "C" cell 1.5v batteries, 2.) A.C. Line Cord (converter is built in), or 3.) Auto cigarette lighter adaptor. This unit features AUTOMATIC LEVEL CONTROL (ALC) and BUILT-IN MIKE for recording and AUTOMATIC END-OF-TAPE SHUTOFF for shutting the tape recorder off when end of tape is reached during PLAY and RECORD modes.

The following CONTROL & JACK functions are incorporated in this model:

"Piano key" Function Buttons – for PLAY, REWIND, FAST FORWARD, RECORD, STOP and EJECT
"Rotary Thumbwheel" type VOLUME control – for PLAYBACK only
Two pin MIC jack – for microphone & patchcord inputs. Also used with ERASE PLUG to de-activate built-in mike when erasing a tape
EAR jack – for earphone & external speaker outputs
PWR jack – for Auto Lighter Adaptor

SERVICE SPECIFICATIONS

ELECTRICAL SPECS

POWER SUPPLY – 7.5vDC
CURRENT DRAINS – With C-60 Cassette installed
- Play Mode – 100 milliamps maximum
- Record Mode – 130 milliamps maximum
- Rewind Mode – 165 milliamps maximum
- Fast Forward Mode – 160 milliamps maximum

SIGNAL TO NOISE RATIO – -25 db minimum
ERASED SIGNAL RATIO – -40 db minimum
CROSSTALK – -40 minimum

MECHANICAL SPECS

TAPE SPEED – 1 7/8 ips $\pm$3% from 5.5 to 7.5vDC
WOW & FLUTTER – Less than .35% RMS
TAKE UP TORQUE – 35 gm/cm At Take Up Hub–minimum
REWIND TORQUE – 50 gm/cm At Supply Hub–minimum
BACKDRAG TORQUE – 0-4 gm/cm At Take Up Hub–maximum
FAST FORWARD TORQUE – 50 gm/cm At Take Up Hub–minimum
CASSETTE WINDING TIME – With C-60 Cassette
- Fast Forward – 2 min. Maximum
- Rewind – 2 min. Maximum

DISASSEMBLY PROCEDURE

I. CABINET BACK REMOVAL

A. Remove one (1) screw from center of cabinet back
B. Remove two (2) rubber feet from handle end of cabinet back.
C. Remove two (2) screws from under rubber feet.
D. Lift cabinet back up and away by disengaging snap tab at A.C. cord socket.

II. TRANSPORT REMOVAL

A. Remove one (1) screw on outside motor tab.
B. Remove one (1) screw from edge of transport plate near flywheel.
C. Remove one (1) screw from edge of transport plate below cutaway on edge of circuit board.
D. Remove one (1) screw from edge of transport plate next to cabinet back screw boss.
E. Lift transport up and out of cabinet front.

NOTE: If transport appears to be "hung up" push the EJECT button to free cassette door assembly.

TAPE HEADS – ALIGNMENT, CLEANING & DEMAGNETIZING

1. ALIGNMENT – Record/Play & Erase heads should be re-aligned whenever the heads are replaced or repaired. The Erase head is aligned by moving its mount until the tape rides directly between the tape guides on the head. The Record/Play head is aligned by peaking it for maximum using the RT7205 (Hartak Mark XII) 3 & 6KHz Azimuth, Flutter and Speed Test Tape. Use the 6KHz section of SIDE 2 for head azimuth alignments.

2. CLEANING & DEMAGNETIZING – Record/Play and Erase heads should be cleaned and demagnetized *EVERY* time a tape recorder is serviced.

   A. Cleaning–Heads should be cleaned with a cotton swab and DENATURED alcohol or commercial head cleaning fluid only; head cleaning cassettes are NOT recommended due to their abrasive characteristics which can affect head performance.

   B. Demagnetizing–Heads should be thoroughly demagnetized using the EA2850 Head Demagnetizer or a suitable equivalent.

TAPE SPEED TEST

1. Connect recorder to a 7.5vDC power supply or RT9541 converter (Do Not Use Batteries).
2. Insert the RT7205 (Hartak Mark XII) Test Cassette in the machine with SIDE 1 up and put machine in PLAY mode.
3. Listen to the tone beeps for a moment then pick a tone beep to start timing sequence.
4. Using a stop watch or sweep second hand on a watch, time the next twelve tone beeps (not counting the starting beep). Repeat 3 times.
5. Check your timings against the following chart to determine if the machine speed is within the standard speed specification of 1 7/8 ips ±3%.

Time	*Speed Factor*	
58.2 seconds	-3%	Slow
58.8 seconds	-2%	Slow
59.4 seconds	-1%	Slow
60 seconds	±0%	Correct
60.6 seconds	+1%	Fast
61.2 seconds	+2%	Fast
61.8 seconds	+3%	Fast

6. Clean all drive surfaces, check & adjust torques, and adjust motor speed (through slot in bottom of motor) if speed variation is more than ±3% (2 seconds).

TORQUE MEASUREMENTS

If a Gram-Centimeter scale is not available, a feel for the required torque can be developed if a known good machine is used. Place the unit in the various modes and lock the spindles with finger tips to develop a feel for the torques developed. During locked spindle conditions, the drain increases approximately as shown below:

Locked Take-Up Clutch .... 5 to 10 ma
Locked Fast Forward Hub .... 100 to 150 ma
Locked Rewind Hub .... 100 to 150 ma

LUBRICATION

The mechanical parts are factory coated with a thin coat of light grease and should not necessarily require further lubrication. If a light grease is applied, use caution not to get any grease on the record and erase heads, hubs, pulleys, drive belts, tape reels or selector switch. Use a good grade of lubricant such as G.E. Silicone Lube G-322L or Lubriplate (00).

MECHANICAL SEQUENCE OF OPERATION

PLAY MODE—When the PLAY pushbutton is depressed, its activating arm comes in contact with the slide base drive plate (19) and the brake arm (20) moving it forward. As the brake arm (20) moves forward, it breaks contact with the tape hubs (5, 6) releasing the "brakes", makes contact with the power switch SW-2 (406) pushing it closed which activates the motor (407), and the amplifier.

At the same time, the slide base assy (2) is moving forward to a latched position. This allows the tape heads (401, 402), pinchwheel assy (12) and the Auto Stop sensor (33) to penetrate into the cassette. The tab on the left front side of the slide base assy (2) comes in contact with the idler assy (24) disengaging it from the flywheel (10). The AUTO STOP trip arm (31) is moved closer to the take-up clutch (5) and is in a retracted position. The right front tip of the slide base assy (2) moves forward allowing the take-up idler (4) to contact the take-up clutch (5).

When SW-2 (406) is closed, voltage is applied to the motor (407). The motor (407) drives the flywheel and capstan assy (10) by means of drive belt (29). The capstan then drives the pinchwheel (12), driving the tape. The drive belt (29) also drives the take-up idler which in turn drives the take-up clutch assy (5).

NOTE: It is important to note that the take-up mechanism is different on this model. The clutch action is provided by the take-up hub instead of the take-up idler (as in M8300 etc.). This is due to the need for the higher torque level required by the AUTO STOP mechanism.

RECORD MODE—When a cassette (with interlock tabs intact) is inserted in the machine, the interlock lever tip comes in contact with the cassette tab causing the interlock lever (21) to be tilted back far enough for the record lever assy (9) to slide forward underneath it when the RECORD pushbutton is depressed. The tab on the lever assy (9) pushes the RECORD/PLAY switch into the RECORD position. The rest of the sequence is identical to the PLAY mode.

FAST FORWARD MODE—When the FAST FORWARD pushbutton is depressed, its activating arm comes in contact with the brake arm (20) moving it forward to release the "brakes" and close the power switch SW-2 (406). At the same time, the idler arm assy (3, 24) is pulled against the flywheel and the take-up clutch (lower half) by spring (38). This allows the take-up hub to drive the tape at a high speed for FAST FORWARD.

REWIND MODE—When the REWIND pushbutton is depressed its activating arm comes in contact with the brake arm (20) moving it forward to release the "brakes" and close SW-2 (406) the power switch. The rewind pulley assy (7, 25) is pulled against the idler assy (24) and the supply hub (6) by the spring (35). This allows the supply hub to drive the tape at a high speed for REWIND.

MANUAL STOP MODE—When the STOP pushbutton is depressed the latch plate for the pushbutton assy (18) is disengaged returning the slide base assy (2), brake arm (20) and all other pulleys, etc., to their original positions.

AUTO STOP MODE—As the tape reaches the end of one side of play, it becomes taut. Since the AUTO STOP sensor (33) is riding against the tape, it senses the increase in tape tension and retracts. As the sensor arm (30) retracts, it pushes the AUTO STOP trip arm (31) forward toward the take-up clutch (5) which is still turning. The raised cam blocks on the clutch (upper hub) hit against the trip arm (31) moving it to the right. This causes the latch plate for the pushbutton assy (18) to disengage, returning all moving parts to their original positions.

NOTE: The AUTO STOP mechanism operates only in PLAY and RECORD modes. It will not operate in REWIND or FAST FORWARD modes because the slide base assy (2) is not in a forward position, therefore, the tape cannot activate the sensor (33).

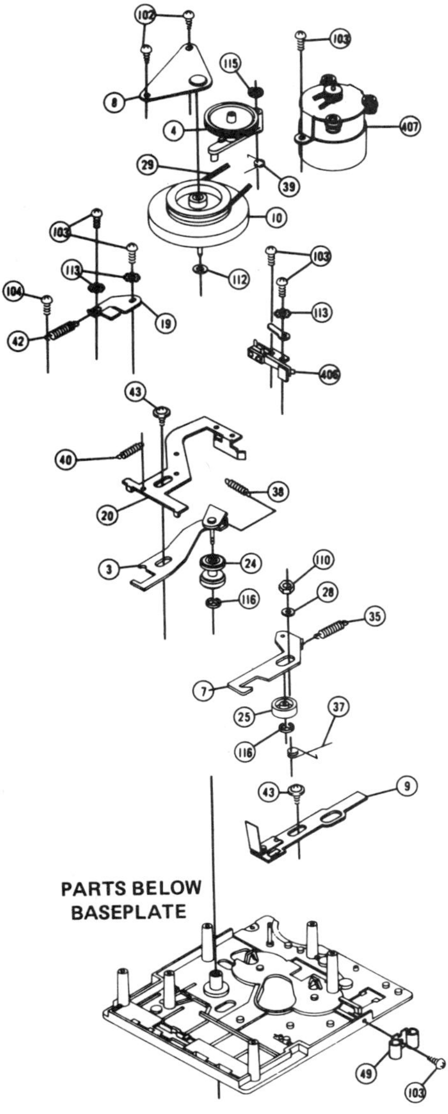

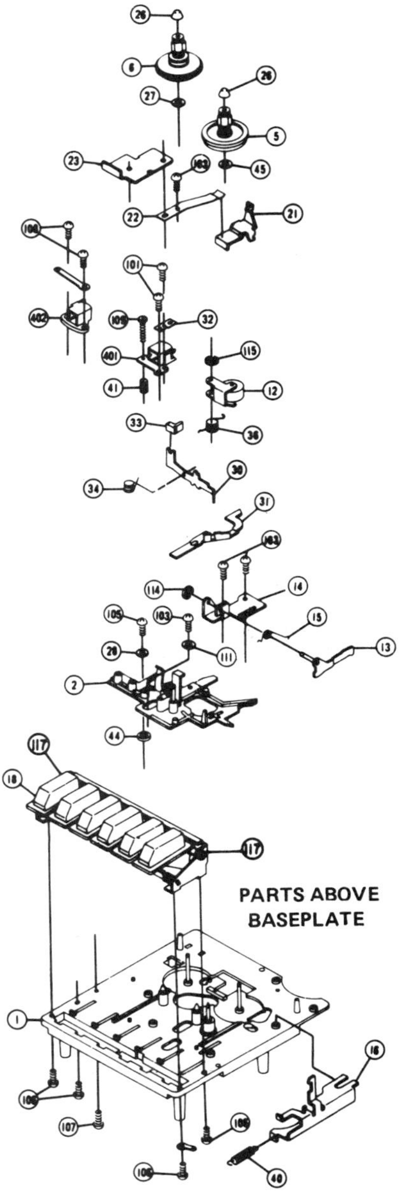

EXPLODED VIEW
MODEL M8445A

MECHANICAL TROUBLESHOOTING CHART

CHECK FOR / SYMPTOM	NO REWIND	NO FAST FORWARD	NO TAKE UP	NO AUTO STOP	NO RECORD LOCK	PUSHBUTTONS WILL NOT LATCH	PREMATURE AUTO STOP	INT REWIND	INT FAST FORWARD	INT TAKE UP	WOW & FLUTTER	TAPE SPILL	CASSETTE WONT EJECT
Pinchwheel pressure too high or low			X	X						X	X		
Take-up torque too high or low			X							X		X	
Rewind torque too high or low	X							X					
Fast Forward torque too high or low		X							X				
Backdrag torque too high or low	X							X				X	
Auto Stop sensor tension too high or low				X			X						
Take-up idler tension too high or low			X	X			X			X		X	
Defective motor (407)	X	X	X								X		
Drive belt (29) broken, worn or dirty	X	X	X					X	X	X	X		
Flywheel & capstan (10) binding, bent or loose	X	X	X					X	X	X	X		
Take-up clutch (5) binding or slipping	X	X	X					X	X	X		X	
Supply hub (6) binding or wobbly	X	X					X	X	X		X		
Rewind pulley (25) binding or wobbly	X							X					
Take-up idler (4) binding or wobbly			X	X						X		X	
Idler (24) binding or wobbly	X	X						X	X				
Pinchwheel (12) worn, wobbly or dirty			X							X	X		
Pushbutton latch plate (18) bent	X	X	X		X	X		X	X	X			
Slide base (2) binding, bent or broken			X	X	X	X	X			X	X	X	
Eject lever (13) bent or broken													X
Rewind pulley arm (7) bent or broken	X							X					
Idler pulley arm (3) bent or broken	X	X						X	X				
Record lock lever (9)bent or broken					X	X							
Interlock lever (21) bent or loose					X	X							
Auto Stop sensor arm (30) bent or loose				X			X						
Auto Stop sensor (33) broken or missing				X									
Auto Stop trip arm (31) broken, bent or loose				X									
Auto Stop cam (on take-up clutch) worn or broken				X									
Brake arm (20) bent or broken	X	X	X			X						X	
Cassette guides bent, loose or burred							X	X	X		X		
Tape heads (401, 402) bent or loose			X				X			X	X		
Defective cassette	X	X	X	X	X		X	X	X	X	X	X	X

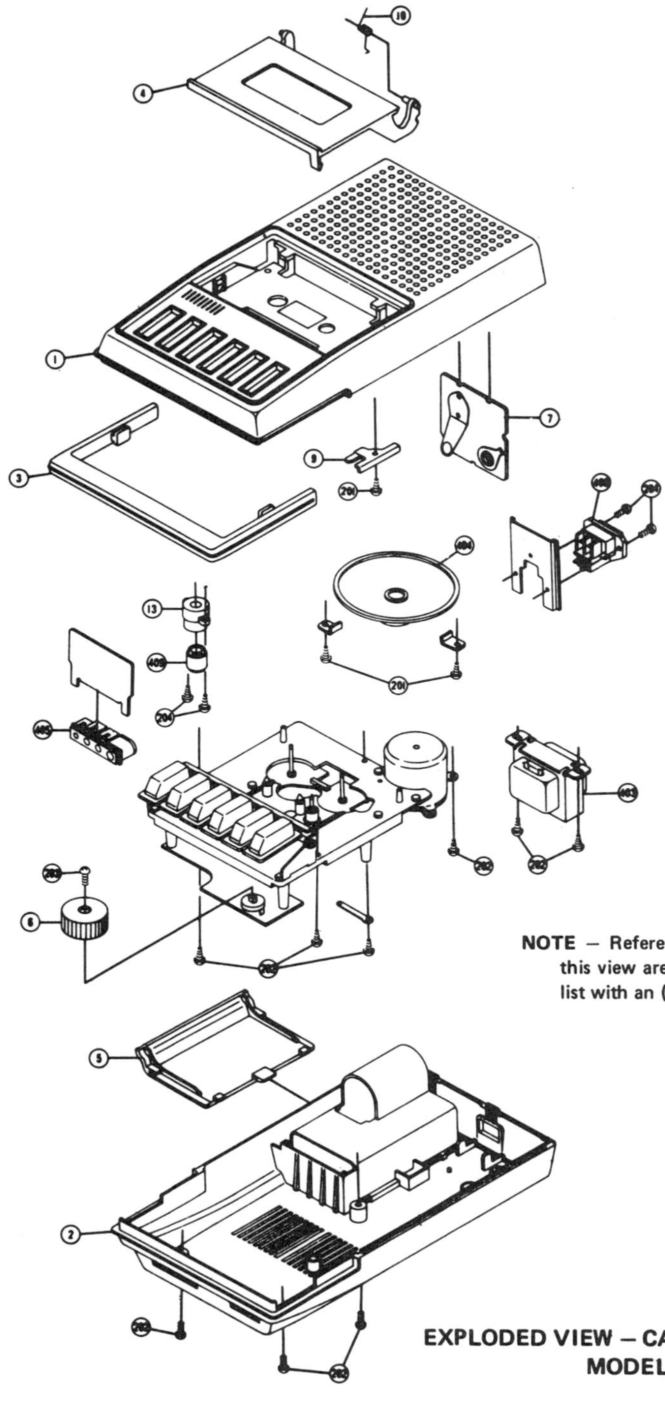

NOTE – Reference numbers from this view are listed in the parts list with an (A) suffix.

EXPLODED VIEW – CABINET & APPEARANCE
MODEL M8445A

VOLTAGES

Since the ground currents existing at various points on the printed ground strap may introduce distortion or erroneous readings, it is important to make readings, supply power, or make signal insertions using the proper physical point on the ground strap.

All voltages are negative with respect to ground taken with a 20 KOHM/V meter.

ELECTRICAL OPERATION

PREAMP—The preamp is a standard direct coupled amplifier consisting of Q1 and Q2. The voltage at the emitter of Q2 provides DC bias (through R8) to the base of Q1. Approximately 0.1 milliamp flows through the collector of Q1 and about 1.0 milliamp flows through the collector of Q2. Frequency shaping and compensation is accomplished by the network of C9, R9 and R13 in PLAYBACK. R12 provides the same function in RECORD.

This preamp is basically flat with respect to its frequency response range and has a high gain ratio. A few millivolts at the input will generate several hundred millivolts of output.

DRIVER AMP—Q3 performs the driver function by providing additional voltage gain which is then transformer coupled to the output stage through T-2.

OUTPUT AMP—The output amplifier consists of Q4 and Q5, PNP germanium power transistors in a push-push configuration. Both stages are biased slightly ON (1 to 10 ma) by the bias network (R23, 24, 25, 26 and 27). The collector of Q5 is tied to the midpoint of the bias network and is about 3.75 volts. A bias voltage of about 0.2 volts is developed by the thermistors TH-1 and 2 which decrease in resistance value as the transistor operating current increases.

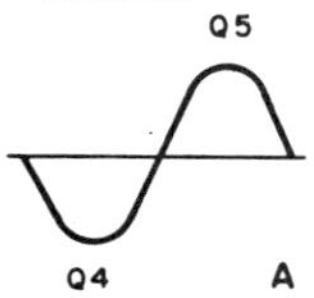

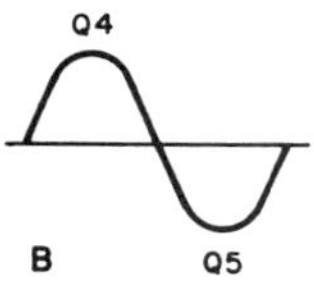

When sine wave A is applied to the output stages, the negative going pulse turns Q4 ON while the positive pulse turns Q5 OFF. Since the voltage on the speaker is also going negative, current flows from the power supply through Q4 and the speaker load to ground.

At the end of the 180° cycle the signal is inverted and sine wave B is applied to Q4 and Q5. The positive pulse turns Q4 OFF while the negative pulse drives Q5 into conduction. No current is taken from the power supply but rather is supplied by the speaker coupling capacitor C19 (220 MFD) as it discharges through Q5 to ground. AC gain stability is provided by feedback to the emitter of Q3 through R28.

ALC CIRCUIT—Because all input or "source" signals vary in amplitude, the ALC circuit (Q6, C12, R14, 30, 34 and D1) acts to minimize these variations so that the recording plays back at a constant output level, regardless of changes in input or "source" level. A sample of the RECORD signal is fed to D1 (through R30) where it is rectified and applied to the base of Q6 as a DC bias which effectively turns Q6 ON and OFF. Since the collector of Q6 is connected to the base of Q1 (preamp input), Q6 will act as an AC shunt. As changes in signal level at Q6 increase and decrease, Q6 will shunt more or less signal at the preamp input. This action tends to prevent Q1 from being driven into distortion and provides stronger ALC action.

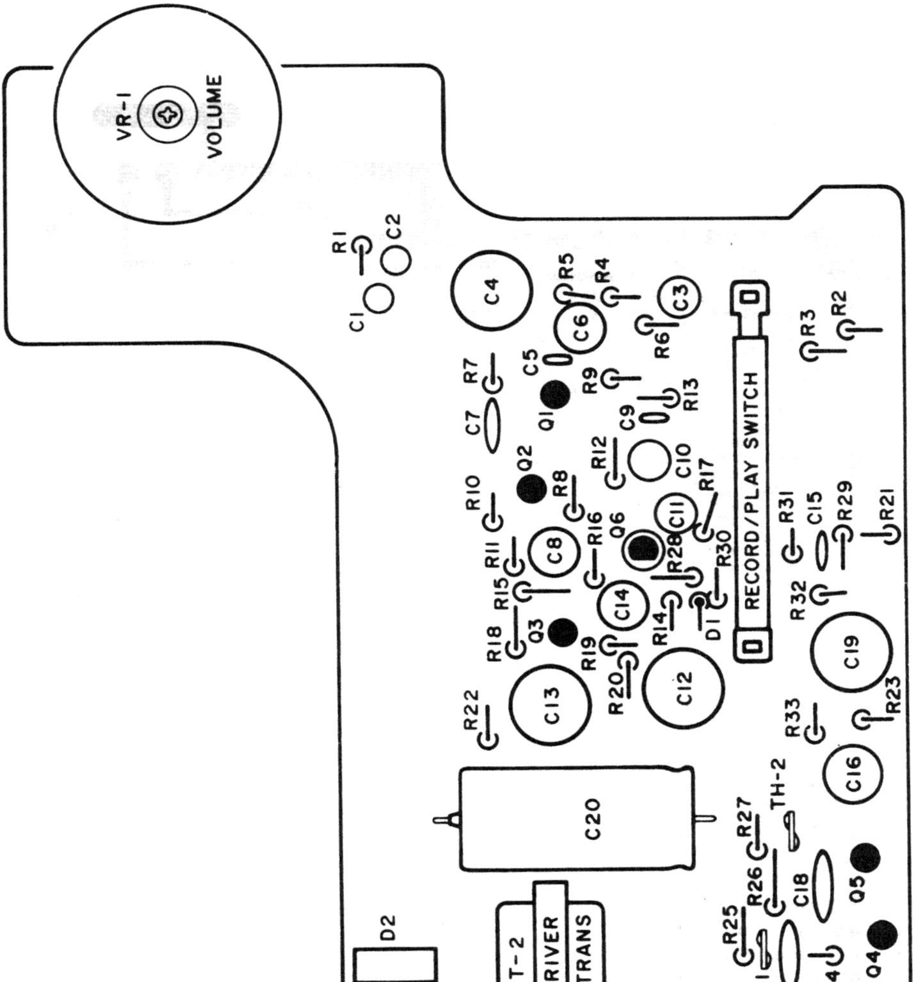
VR-1
VOLUME
RECORD/PLAY SWITCH
T-2
DRIVER
TRANS
C20
D2

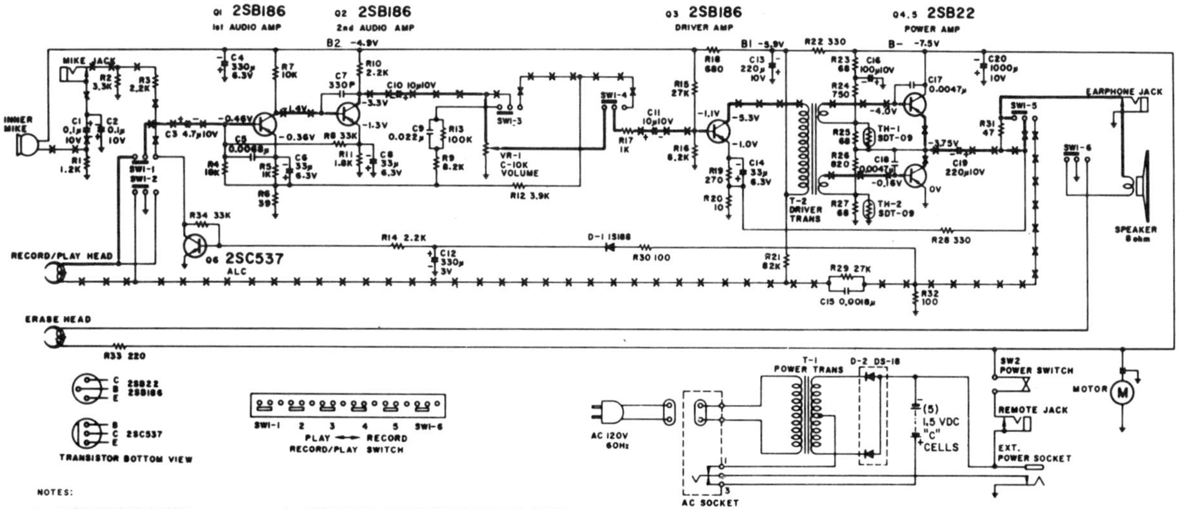

NOTES:

1. UNLESS OTHERWISE NOTED
   CAPACITORS MORE THAN 1 = PF
   CAPACITORS LESS THAN 1 = MF
   RESISTORS IN OHMS, K = 1,000
   M = 1,000,000

2. VOLTAGES ARE NEGATIVE WITH RESPECT TO GROUND TAKEN IN PLAYBACK POSITION EXCEPT FOR Q6 WHICH IS TAKEN IN RECORD POSITION

3. SIGNAL PATHS ARE AS FOLLOWS
   RECORD —×—×—
   PLAYBACK ————
   RECORD AND PLAYBACK —✱—✱—

SCHEMATIC DIAGRAM
MODEL M8445A

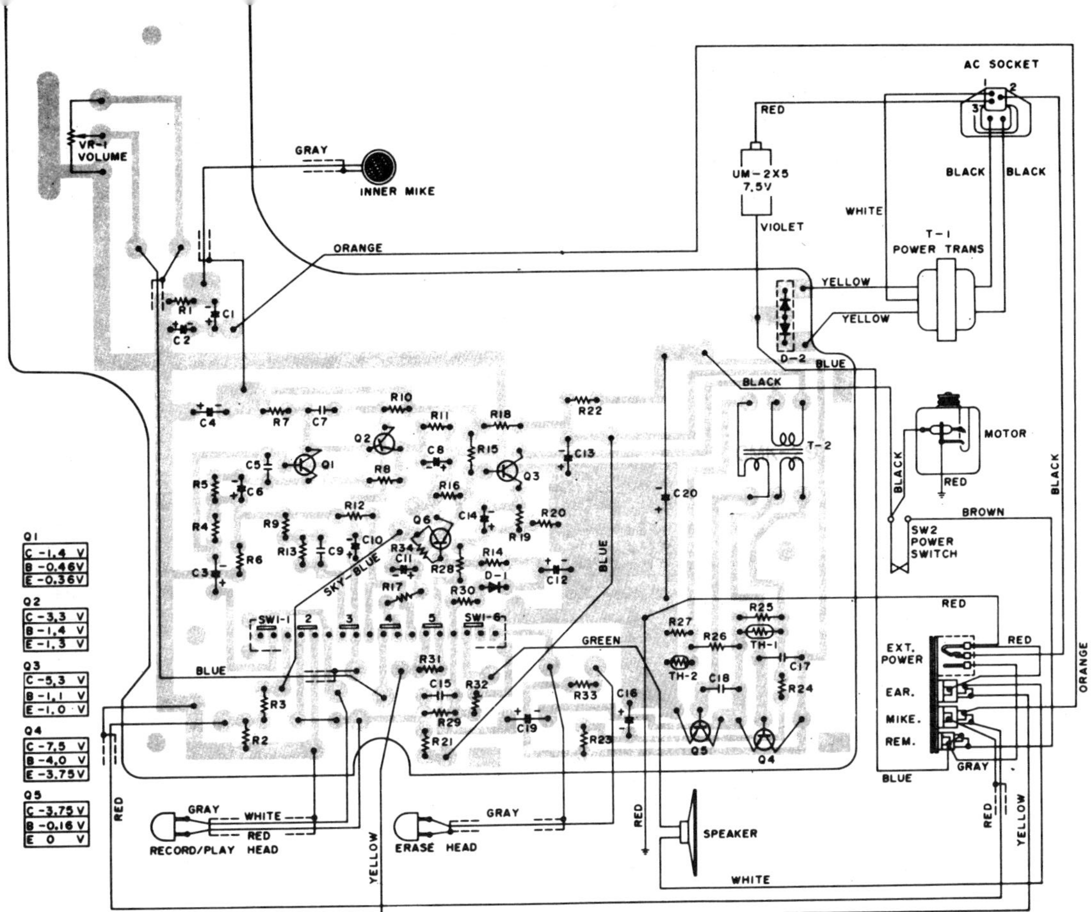

Q1	Q2	Q3	Q4	Q5
C -1.4 V	C -3.3 V	C -5.3 V	C -7.5 V	C -3.75 V
B -0.46V	B -1.4 V	B -1.1 V	B -4.0 V	B -0.16 V
E -0.36V	E -1.3 V	E -1.0 V	E -3.75V	E 0 V

WIRING DIAGRAM – BOTTOM VIEW
MODEL M8445A

ELECTRICAL TROUBLESHOOTING CHART

CHECK FOR \ SYMPTOM	NO PLAYBACK	WEAK PLAYBACK	DISTORTED PLAYBACK	NO RECORD	WEAK RECORD	DISTORTED RECORD	NO ERASE	PARTIAL ERASE	NO ALC ACTION
Q1 open or shorted	X		X	X		X			
Q2 open or shorted	X		X	X		X			
Q3 open or shorted	X		X	X		X			
Q4 open or shorted	X		X	X		X			
Q5 open or shorted	X		X	X		X			
Q6 open or shorted				X	X				X
D1 open or shorted					X	X			X
C1 open or shorted				X	X	X			
C3 open or shorted	X	X		X	X				
C6 open or shorted		X			X				
C8 open or shorted		X			X				
C10 open or shorted	X	X		X	X				
C11 open or shorted	X	X		X	X				
C12 open or shorted					X				X
C14 open or shorted		X			X				
C19 open or shorted	X	X	X	X	X	X			
T1 open or shorted	X	X	X	X	X	X	X	X	X
T2 open or shorted	X	X		X	X				
Record/Play head open or shorted	X	X	X	X	X	X			
Erase head open or shorted						X	X	X	
Speaker open or shorted	X	X	X						
Ear jack open	X								
Mike jack open				X					
Record/Play switch defective	X			X			X		X
No or low record bias (.02vDC at head)				X	X	X			
No or low erase voltage (4vDC at head)						X	X	X	
No or low B+ voltage (7.5vDC at SW-2)	X	X	X	X	X	X	X	X	X
Heads wired in wrong polarity						X		X	
Heads dirty or magnetized		X	X		X	X		X	
Heads bent or misaligned		X	X		X	X		X	
Defective cassette		X	X		X	X		X	
Volume Control open or shorted	X	X	X						

REPLACEMENT PARTS LIST
MECHANICAL PARTS

CATALOG NUMBER	ITEM NUMBER	DESCRIPTION
*EA92X7	1	Chassy Plate Assy
RV1238	2	Slide Base Assy
RV1243	3	Idler Arm Assy
RV1239	4	Take-Up Idler Assy
RV1240	5	Take-Up Clutch Assy
RV1241	6	Supply Hub Assy
RV1242	7	Rewind Pulley Arm Assy
RV1244	8	Flywheel Support Assy
RV1246	9	Record Lever Assy
RV1254	10	Flywheel & Capstan Assy
RV1960	12	Pinchwheel Assy
*EA5X16	13	Eject Lever Assy
*EA2X117	14	Eject Lever Bracket
*EA3X72	15	Eject Lever Spring
*EA2X118	16	Cassette Door Lock Arm
*EA92X8	18	Pushbutton Assy w/o Buttons
RV1269	19	Slide Base Drive Plate
RV1251	20	Brake Arm
*EA2X141	21	Interlock Lever
RV1261	22	Interlock Lever Spring
*EA2X119	23	Chassis Mtg. Plate
RV1255	24	Idler Roller
RV1256	25	Rewind Roller
RT7913	26	Reel Cap
RV1259	27	Spindle Washer
RT5753	28	Arm Spacer
RT2236	29	Drive Belt
RV1249	30	Auto Stop Sensor Arm
RV1966	31	Auto Stop Trip Arm
RV1253	32	Stop Lever Guide Plate
RV1258	33	Auto Stop Sensor
RV1267	34	Auto Stop Sensor Arm Spring

CATALOG NUMBER	ITEM NUMBER	DESCRIPTION
RT5751	35	Rewind Roller Spring
RV1263	36	Pinchwheel Spring
RV1268	37	Record Plate Spring
RV1265	38	Idler Arm Spring
RV1266	39	Take-Up Idler Spring
RV1262	40	Brake Arm Spring
RV1264	41	Azimuth Adjust Spring
RV1270	42	Slide Base Spring
RV1969	43	Pivot Screw w/shoulder
*EA1X54	44	Slide Base Spacer
*EA1X84	45	Hub Washer
RT6311	49	Heat Sink
RV2185(k)	101	Screw, Pan Hd. Tapping 2 x 6 mm #2
RV2185(k)	102, 201	Screw, Pan Hd. Tapping 3 x 8 mm #2
RV2185(k)	103	Screw, Pan Hd. Machine 2.5 x 6 mm
RV2185(k)	104	Screw, Pan Hd. Machine 2.5 x 8 mm
RV2185(k)	105	Screw, Pan Hd. Machine 2.5 x 12 mm
RV2185(k)	106	Screw, Pan Hd. Machine 2.6 x 6 mm
RV2185(k)	107	Screw, Pan Hd. Machine 3 x .5 x 6 mm
RV2185(k)	108	Screw, Truss Hd. Tapping 2 x 6 mm #1
RV2185(k)	109	Screw, Cylinder Hd. Machine 2 x 10 mm
RT3602(k)	110	Nut, 2.6 mm
RV2185(k)	111	Washer, 2.9 x 7.5 x .5 mm
RV2185(k)	112	Washer, 2 x 7 x .3 mm
RV2185(k)	113	Washer, Startooth 2.6 mm
RT6288	114	"E" Ring 2.5 mm
RT6667	115	"E" Ring 2 mm
RT6666	116	"E" Ring 1.5 mm
RT6669	117	"E" Ring 3 mm
RV2185(k)	202	Screw, Pan Hd. Tapping 3 x 12 mm #2
RV2185(k)	203	Screw, Truss Hd. Machine 2 x 6 mm
RV2185(k)	204	Screw, Pan Hd. Tapping 2.6 x 8 mm #2

* – Indicates new part not previously cataloged

(k) – Indicates item is part of a kit

NOTE–Item 44 used on some sets only. It does not have to be added to assy if not previously installed.

REPLACEMENT PARTS LIST

CATALOG NUMBER	SYMBOL	DESCRIPTION	
	CAPACITORS		
RT5264	C1, 2	.1mfd, 10v	(A)
RT4933	C3	4.7mfd, 10v	(E)
RT7632	C4	330mfd, 6.3v	(E)
RT5438	C5	.0068mfd, 50v	(M)
RT5269	C6, 8, 14	33mfd, 6.3v	(E)
RT4626	C7	330pf, 50v	(C)
RT4563	C9	.022mfd, 50v	(M)
RT2306	C10, 11	10mfd, 10v	(E)
RT5777	C12	330mfd, 3v	(E)
RT5047	C13, 19	220mfd, 10v	(E)
RT5774	C15	.0018mfd, 50v	(C)
RT6099	C16	100mfd, 10v	(E)
RT2860	C17, 18	.0047mfd, 50v	(C)
EA2159	C20	1000mfd, 10v	(E)

(A) ALSICON (C) CERAMIC
(E) ELECTROLYTIC (M) MYLAR

RESISTORS & POTENTIOMETERS

CATALOG NUMBER	SYMBOL	DESCRIPTION
RT9730(k)	R1	1.2K
RT9730(k)	R2	3.3K
RT9730(k)	R3, 10, 14	2.2K
RT9730(k)	R4	18K
RT9730(k)	R5, 17	1K
RT8167	R6	39 ohm
RT9730(k)	R7	10K
RT9730(k)	R8, 34	33K
RT9730(k)	R9, 16	8.2K
RT9730(k)	R11	1.8K
RT9730(k)	R12	3.9K
RT9730(k)	R13	100K
RT9730(k)	R15, 29	27K
RT5253	R18	680 ohm
RT9730(k)	R19	270 ohm
RT9730(k)	R20	10 ohm
RT9730(k)	R21	82K
RT5252	R22, 28	330 ohm
RT2319	R23, 25, 27	68 ohm
RT8006(k)	R24	750 ohm, ¼ Watt 5%
RT8006(k)	R26	820 ohm
RT9730(k)	R30, 32	100 ohm
RT9730(k)	R31	47 ohm
RT9730(k)	R33	220 ohm
*EA49X179	VR1	Volume Control, 10K
RT7635	TH-1, 2	Thermistor, SDT-09

NOTE—All resistors are ¼ Watt @ 10% unless noted

COILS & TRANSFORMERS

CATALOG NUMBER	SYMBOL	DESCRIPTION
RV1278	T1 (403)	Power Transformer
RV2256	T2	Driver Transformer

* Denotes parts not previously cataloged
(k) Indicates item is part of a kit

CATALOG NUMBER	SYMBOL	DESCRIPTION
	TRANSISTORS & DIODES	
RT2330	Q1	1st Audio Amp, 2SB186
RT2330	Q2	2nd Audio Amp, 2SB186
RT2330	Q3	Driver Amp, 2SB186
RT2331	Q4, 5	Power Amp, 2SB22
RT4761	Q6	ALC Amp, 2SC537
RT2334	D1	ALC Diode, 1S188AM
RV2289	D2	Rectifier Diode, DS-18

CABINET & APPEARANCE ITEMS

CATALOG NUMBER	SYMBOL	DESCRIPTION
*EA98X46	1A	Cabinet Top Assy
*EA98X47	2A	Cabinet Bottom
*EA78X21	3A	Handle
*EA9X27	4A	Cassette Door
*EA9X25	5A	Battery Door
*EA43X242	6A	Volume Knob
*EA43X243	17A	Pushbutton—Black
*EA43X244	18A	Pushbutton—Blue
*EA45X4	19A	Cabinet Foot

MISCELLANEOUS

CATALOG NUMBER	SYMBOL	DESCRIPTION
*EA2X114	7A	Battery Terminal Plate Assy
*EA2X115	9A	Cassette Door Bracket
*EA3X70	10A	Cassette Door Spring
*EA9X26	13A	Microphone Holder
RV1237	14A	Battery Terminal
*EA3X71	15A	Cassette Tension Spring
*EA2X116	16A	Battery Terminal
RV1979	401	Record/Play Head
RV1980	402	Erase Head
RB2515	404	Speaker 8 ohm
*EA67X72	405	Jack Assy
RV1276	406	Power Switch (SW-2)
*EA72X1	407	Motor
*EA34X40	408	A.C. Power Socket
*EA62X84	409	Built-In Mike
RV1988	SW-1	Record/Play Switch
*EA66X8		A.C. Power Cord
*EA93X141		Amplifier P.C. Board w/comp

ACCESSORY ITEMS

CATALOG NUMBER	SYMBOL	DESCRIPTION
RV1283		External Microphone, 600 ohm
RV1284		Microphone Stand
RT7597		Earphone, 8 ohm
RT9508		Auto Lighter Adaptor
RT9607		Patchcord Kit
RT9057		Footpedal Remote Switch
RT9546		Hand Remote Switch
RV2287		Erase Plug
RT7201		C-60 Cassette (Blank)
RT7220		C-120 Cassette (Blank)
RT7230		Head Cleaning Cassette

(A) Indicates item on exploded view of cabinet

MODELS
M8700A, B, C
"MINI"

TAPE
RECORDER

FILE - M

GENERAL ELECTRIC
SERVICE INFORMATION

CAUTION: THIS MANUAL IS DESIGNED FOR USE BY QUALIFIED ELECTRONIC TECHNICIANS ONLY. CONSUMER USERS ARE URGED TO CONTACT QUALIFIED FACTORY AUTHORIZED SERVICE FACILITIES FOR REPAIRS.

GENERAL SPECIFICATIONS

GENERAL SPECIFICATIONS
POWER OPTIONS DC - RT9543 AC Convertor DC - (5) 1.5vDC "AA" Cell Batteries DC - RT9508 Auto Lighter Adaptor
TAPE SPEED - 1-7/8 i.p.s. $^{+3\%}_{-4\%}$ from 5.5 to 7.5vDC
IMPEDANCES INPUT - Approx. 600 ohms OUTPUT - 8 ohms EXT. MIKE - 600 ohms EARPHONE - 8 ohms SPEAKER - 8 ohms PATCHCORD MATCHING - 200 to 600 ohms
FREQUENCY RESPONSE PRE-RECORDED CASSETTE - 200 to 6KHz ±5db MIKE RECORDED CASSETTE - 200 to 4KHz ±5db
POWER OUTPUT - 400 mw min. at 10% THD
RECORD BIAS - AC, appx. 511KHz
ERASE VOLTAGE - DC (A & B), AC on "C" version

GENERAL INFORMATION

General Electric Models M8700A, B and C are "MINI" portable cassette tape recorders that are electrically similar to each other. These models incorporate all the standard tape recorder functions in addition to AUTOMATIC LEVEL CONTROL (ALC) and BUILT-IN MICROPHONES for recording. They operate on 7.5vDC which can be supplied by 1) 5"AA" cell 1.5v batteries, or 2) RT9543 AC converter. The "B" & "C" versions have a combination PAUSE/BATTERY CONDITION push-button that stops the motor when depressed slightly and engages the battery condition indicator when depressed further. The "A" version has a dynamic mike element while the "B & V" versions have condenser mikes. The "A & B" versions use DC ERASE while the "C" version uses AC ERASE.

Due to the extend of miniaturization and compactness inherent in these units it is necessary to be extremely careful when repairing or replacing parts so that damage to other parts and components does not occur. It is imperative that all applicable adjustments and/or alignments be performed with the specified tools and equipment each time service is performed on unit.

SERVICE SPECIFICATIONS

ELECTRICAL SPEC'S

1. POWER SUPPLY - 7.5v DC at 250 ma

2. CURRENT DRAINS w/C-60

   A. Play - 120ma max., 87ma Ave.
   B. Record - 150ma max., 109ma Ave.
   C. Fast Forward - 150ma max., 114ma Ave.
   D. Rewind - 150ma max., 100ma Ave.

3. SIGNAL TO NOISE RATIO - -30db minimum

4. ERASED SIGNAL RATIO - -50db minimum

5. CROSS TALK - better than - 40db

6. RECORD BIAS (Across 10ohms) - See applicable voltage view

7. ERASE VOLTAGE - See applicable voltage view

8. BIAS OSC. OUTPUT - See applicable voltage view

9. RECORD HEAD VOLTAGE - See applicable voltage view

MECHANICAL SPEC'S

1. TAPE SPEED - 1-7/8 i.p.s. $^{+3\%}_{-4\%}$ from 7.5 to 4.5v DC

2. WOW & FLUTTER - less than 0.60% RMS

3. TORQUES - w/Cortex guage

   A. Play - 30-60gm/cm
   B. Rewind - 50gm/cm minimum
   C. Fast Forward - 50gm/cm minimum
   D. Back Drag (Rewind) - 0-4gm/cm

4. CASSETTE WINDING TIME w/c-60

   A. Rewind - 180 seconds
   B. Fast Forward - 180 seconds

DISASSEMBLY

A. GENERAL INSTRUCTIONS

Due to the compactness of these units the hardware items removed during disassembly are quite small and easy to lose. It is suggested that these parts be placed in a suitable container during service and that particular attention be paid to both the size of the screws, etc. and their respective locations. This disassembly procedure is arranged in a chronological sequence so that the unit is disassembled only far enough to make the required repair or adjustment. It is suggested that each step be followed in sequence to avoid removal of unnecessary parts. All numbers in ( ) refer to the chassis and cabinet exploded views contained in this manual.

B. CABINET REMOVAL

1. Remove two screws each from the cabinet top (7) and cabinet bottom (9).

2. Remove two screws from cabinet side B (8)

3. Remove control knobs and one screw from cabinet side A (2) only if necessary to gain access to motor or microphone.

4. Remove four screws from chassis panel (10)

Repairs and adjustments to the following can be performed at this point - motor governor, pinchwheel assy, tape heads, eject mechanism, microphone, motor, speaker and bottom of circuit board.

C. CIRCUIT BOARD REMOVAL

1. Remove three screws on outer edges of board

2. Lift speaker out of its well - Note: Rewind hub assy (83) is now accessable

3. It is not necessary to completely remove the circuit board in order to replace components or make adjustments. Lift the circuit board up only as far as is required to reach the desired component on test point. Make sure that any wires or cables moved during repair are properly re-positioned after repair is completed.

D. MOTOR REMOVAL

1. Follow steps 1-4 under CABINET REMOVAL

2. Place unit in PLAY mode and locate the motor mounting screws that become visible when the pinchwheel Assy (63) moves forward.

3. Remove these two screws and slide right hand motor bracket (49) off the motor and remove motor.

E. FLYWHEEL & DRIVE BELT REMOVAL

1. Follow steps 1-4 under CABINET REMOVAL

2. Follow steps 1-3 under CIRCUIT BOARD REMOVAL

3. Remove three screws on capstan bearing plate (33) and remove plate

4. Gently pull upon flywheel until drive belt (29) and bearing (25) are visible

5. To replace flywheel simply remove old one and install new one making sure drive belt, spacers and bearing are properly seated

6. To replace drive belt remove it from flywheel, lift circuit board up and remove it from the take-up clutch and idler pulley. Clean flywheel drive surfaces whenever blet is replaced.

7. Perform flywheel and motor tension adjustments (See ADJUSTMENTS)

MECHANICAL ADJUSTMENTS

FLYWHEEL ADJUSTMENT (Refer to Fig. 1)

1. Supply 7.5vDC to unit, from an external power supply

2. Place unit in PLAY mode and observe "no signal" current drain. If drain is less than 75 milliamps do not adjust flywheel. If the drain exceeds 75 milliamps procede as follows.

3. Loosen the three capstan bearing plate screws and adjust the position of the plate until it provides a current drain of 75 milliamps maximum

4. Tighten plate down, check current drain and readjust if necessary

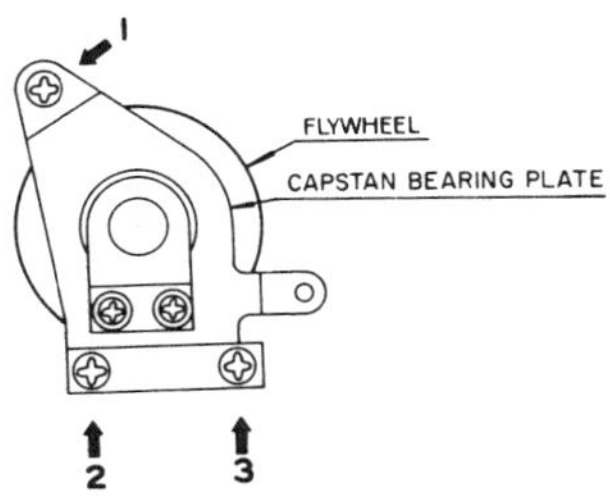

FIGURE I

MOTOR PRESSURE ADJUSTMENT (Refer to Fig. 2)

1. Insert the RT7205 (Hartak Mark XII) test cassette in the unit with the 3KHz section of side 2 up

2. Place unit in play mode and listen to playback for wow & flutter

3. Rotate adjusting screw clockwise or counter-clockwise until wow & flutter is minimized. If a wow & flutter meter is available, the reading should be less than 0.55%

4. Proceed to tape speed test

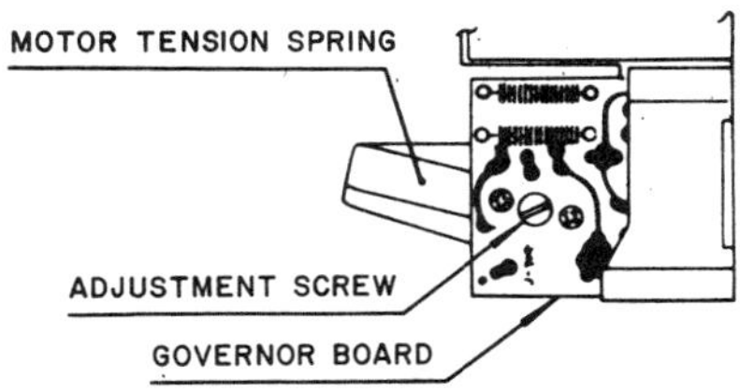

FIGURE 2

TAPE SPEED TEST

1. Connect unit to a 7.5vDC power supply or RT9543 converter (Do not use batteries)

2. Insert the RT7205 (HARTAK Mark XII) Test Cassette in the machine with SIDE 1 up and put unit in PLAY mode

3. Listen to the tone beeps for a moment then pick a tone beep to start timing sequences

4. Using a stop watch or sweep second hand on a watch, time the next twelve tone beeps (not counting the starting beep). Repeat 3 times

5. Check your timings against the following chart to determine if the machine speed is within the speed specification of 1-7/8 i.p.s. +3% -4%

Time	Speed Factor	
57.6 seconds	-4%	Slow
58.2 seconds	-3%	Slow
58.8 seconds	-2%	Slow
59.4 seconds	-1%	Slow
60 seconds	±0%	Correct
60.6 seconds	+1%	Fast
61.2 seconds	+2%	Fast
61.8 seconds	+3%	Fast

6. Adjust motor governor potentiometer until speed is within the specified limits. If a counter is available, the readout should be 2880 to 3090Hz.

TAKE-UP TORQUE ADJUSTMENT (Refer to Fig. 3)

1. Remove three circuit board screws and lift board up and move it toward flywheel

2. Loosen the two retaining screws on the take-up clutch tension spring (45) approximately one turn

3. Slide the spring toward point A to increase torque, toward point B to decrease it, until the proper torque level is shown on the RT9511 Cortex guage

4. Tighten the two screws and recheck the torque reading in PLAY mode. Also check the back drag torque in REWIND mode

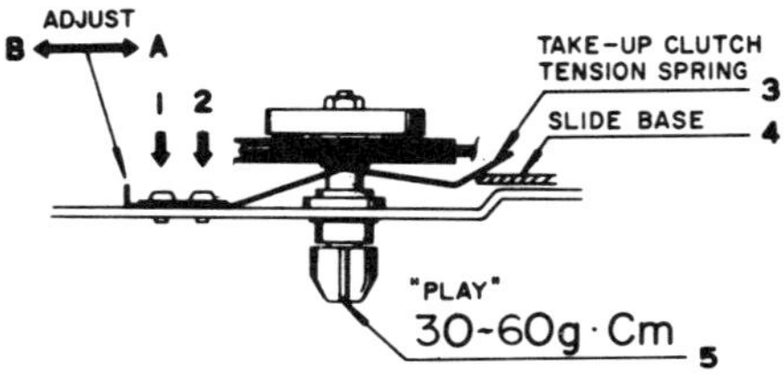

FIGURE 3

FAST FORWARD TORQUE ADJUSTMENT
(Refer to Fig. 4)

1. Remove three circuit board screws and lift board up and move it toward flywheel

2. Clean rubber surface on the fast forward roller (36) and recheck torque. If improper torque still exists proceed to next step

3. Loosen one screw on forward idler arm spring (52) approximately one turn

4. Slide the spring towards point A to increase torque, toward point B to decrease it, until the proper torque level is shown on the RT951 Cortex guage

5. Tighten the screw and recheck torque level

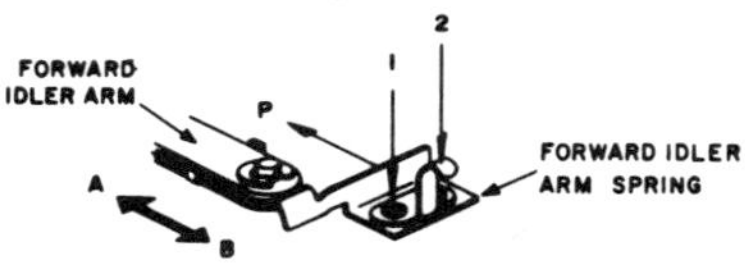

FIGURE 4

REWIND TORQUE ADJUSTMENT

1. The rewind torque on these machines is fixed. It is governed only by the motor force against the flywheel and the condition of the drive belt. If the torque reading is improper, cleaning or replacing the fly-wheel and drive belt will correct it.

PINCHWHEEL PRESSURE (Refer to Fig. 5)

1. If pinchwheel pressure is inadequate to properly drive tape, move the tension to slot B. If this does not correct the pressure replace the pinchwheel assy (63)

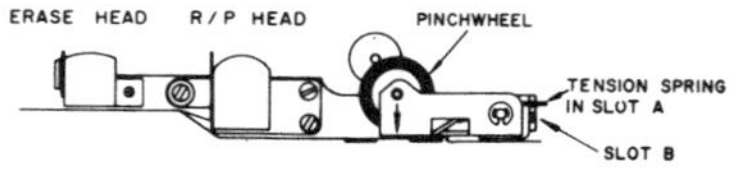

FIGURE 5

ELECTRICAL ADJUSTMENTS

RECORD/PLAY HEAD AZIMUTH ALIGNMENT
(Refer to Fig. 6)

1. Insert the RT7205 (Hartak Mark XII) Test Cassette in the unit with the 6KHz section of side 2 up

2. Place unit in PLAY mode and adjust head azimuth screw through hole in edge of chassis panel. Peak head for maximum audio output

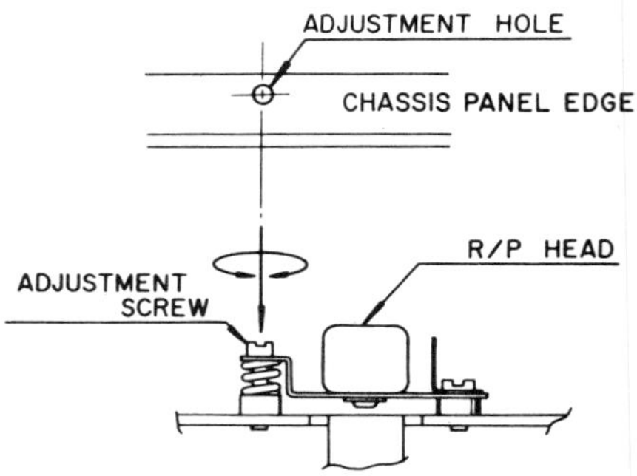

FIGURE 6

BIAS OSCILLATOR ALIGNMENT (Refer to Applicable Voltage View)

1. Set the bias adjust pot to the voltage indicated on the applicable voltage view

2. Peak L-1 for minimum passage of AC on amplifier side

3. Readjust bias pot for voltage across the 10ohm head resistor as specified on the applicable voltage view

4. Place a clean tape in unit and record two signals (600Hz & 6KHz) at - 50db (approximately .003vRMS) three times

5. Play the recordings back and monitor output on an AC VTVM. The outputs should be within 4db of each other

6. Readjust the bias pot until the 4db range is reached

BATTERY CONDITION LAMP ADJUST

1. Connect unit to 7.5vDC power supply (Do not use batteries or converter)

2. Set power supply to 5.5vDC

3. Depress BATT/PAUSE button and adjust lamp pot until the lamp just goes off at 5.5vDC

HEAD CLEANING & DEMAGNETIZING - Record/Play and Erase heads should be cleaned and demagnetized EVERY time a tape recorder is serviced.

A. Cleaning - Heads should be cleaned with a cotton swab and DENATURED alcohol or commercial head cleaning fluid only; head cleaning cassettes are NOT recommended due to their abrasive characteristics which can affect head performance.

B. Demagnetizing - Heads should be thoroughly demagnetized using the EA2850 Head Demagnetizer or a suitable equivalent.

TORQUE MEASUREMENTS

If a Gram-Centimeter scale is not available, a feel for the required torque can be developed if a known good machine is used. Place the unit in the various modes and lock the spindles with finger tips to develop a feel for the torques developed. During locked spindle conditions, the drain increases approximately as shown below:

Locked Take-Up Clutch .... 5 to 10 ma
Locked Fast Forward Hub .... 50 to 100 ma
Locked Rewind Hub .... 75 to 100 ma

LUBRICATION

The mechanical parts are factory coated with a thin coat of light grease and should not necessarily require further lubrication. If a light grease is applied, use caution not to get any grease on the record and erase heads, hubs, pulleys, drive belts, tape reels or selector switch. Use a good grade of lubrication as G. E. Silicone Lube G-322L or Lubriplate (00).

MECHANICAL TROUBLESHOOTING CHART

CHECK FOR / SYMPTOM	NO REWIND	NO FAST FORWARD	NO TAKE UP	NO RECORD LOCK	PUSHBUTTONS WILL NOT LATCH	INT REWIND	INT FAST FORWARD	INT TAKE UP	WOW & FLUTTER	TAPE SPILL	CASSETTE WON'T EJECT
Pinchwheel pressure too high or low			X					X	X		
Take-up torque too high or low			X					X		X	
Rewind torque too high or low	X					X					
Fast Forward torque too high or low		X					X				
Backdrag torque too high or low	X					X				X	
Defective motor (48)	X	X	X						X		
Drive belt (29) broken, worn or dirty	X	X	X			X	X	X	X		
Flywheel & capstan (27) binding, bent or loose	X	X	X			X	X	X	X		
Take-up clutch (30, 32) binding or slipping	X	X	X			X	X	X		X	
Supply hub (83) binding or wobbly	X	X				X	X		X		
Pinchwheel (63) worn, wobbly or dirty			X					X	X		
Slide base (62) binding, bent or broken			X	X	X			X	X	X	
Eject lever (74) bent or broken											X
Rewind pulley arm (39) bent or broken	X					X					
Idler pulley arm (35) bent or broken	X	X				X	X				
Record lever (6) bent or broken				X	X						
Interlock lever (12) bent or loose				X	X						
Cassette guides bent, loose or burred						X	X		X		
Tape heads (68, 69) bent or loose			X					X	X		
Defective cassette	X	X	X	X		X	X	X	X	X	X

ELECTRICAL OPERATION

GENERAL - The following analysis makes reference to component symbols and valves used on the "A" version unit. Some symbol numbers and valves are different on the "B & C" versions.

PREAMP - The preamp, consisting of Q1 and Q2, is a standard direct coupled amplifier. The voltage at the emitter of Q2 provides base bias for Q1 through R1. Signal is coupled in by C3, amplified by Q1 & 2 and fed to the 1st audio Amp & Driver through C8. Flat amplifier response during playback is accomplished by the post-equalization bass boost network of C5, R4 & R5. C7, R6 and R7 provide the same function in RECORD mode.

1st AUDIO AMP - Signal is fed through the volume control (PLAY mode) or through the RIP switch (RECORD mode) and C11 to Q3. The signal is amplified and then direct coupled to the Driver Amp Q4. The voltage at the emitter of Q4 provides base bias for Q3 through R13.

DRIVER AMP - The Driver Amp, Q4, provides additional amplification of the signal which is then direct coupled to the power amp Q5 and Q6.

OUTPUT STAGE - This stage uses an NPN and a PNP power transmitter in a complimentary symmetry configuration. A thermistor, TH-1, is used to provide bias stability. The thermistor value decreases with increasing temperature, thus compensating for transistor changes. A positive going signal, on the base of Q5, turns this transistor ON. At the same time, a negative pulse keeps Q6 OFF. At this time, the voltage on the speaker is also going positive. Thus load current flows from supply through Q5 and the speaker load. The reverse occurs during the next 180 degrees, namely a negative signal cuts off Q5 and a positive Q6 into conduction. During this half cycle, no current is taken from the power supply, but rather is supplied by the 100MFD capacitor (C-15). This circuit then is completed through the speaker coupling capacitor and Q6. AC gain stability and reduction of distortion is provided by feedback through R21, R2 to the emitter of Q3.

ALC CIRCUIT - The ALC function is performed by the circuit of Q7, C19, 21, R24, 29 and D1. When a signal is recorded on tape the level of the signal varies. To eliminate this, a sample of the varying signal is fed to D1 through R29, where it is rectified and applied to the base of Q7 as a DC bias which effectively turns Q7 ON and OFF. Since the collector of Q7 effectively parallels the load of Q1, changes in Q7's collector voltage will cause changes in Q1's output level. This action will tend to eliminate the signal level variations on the tape, allowing a very even recording.

ELECTRONIC MOTOR GOVERNOR - This circuit is basically a voltage regulator. It will maintain a constant motor speed through a range of variations in supply voltage and/or temperature changes. VR-3 is used to adjust for nominal motor speed. Increasing the resistance of VR-3 will increase the base bias voltage on Q10 turning it ON. When Q10 starts conducting, more base current flows through Q9 turning it ON. Turning Q9 ON will cause the motor voltage to increase. Regulation is accomplished by the feedback loop through R36 and D2.

BIAS OSCILLATOR - Q8 is the bias oscillator. This stage provides bias for record head and erase voltage ("C" version only). VR-2 is used to adjust bias current. The trap coil is adjusted to parallel resonate with C17 to minimize bias frequency from getting into the amplifier and maximize bias signal at record head. VR-2 is then adjusted for about 500 microamps in R/P head or about 5mv rms across R39 (10ohms). See Adjustments for proper valves

ELECTRICAL TROUBLESHOOTING CHART

VERSION A	VERSION B	VERSION C	CHECK FOR	SYMPTOM: NO PLAYBACK	WEAK PLAYBACK	DISTORTED PLAYBACK	NO RECORD	WEAK RECORD	DISTORTED RECORD	NO ERASE	PARTIAL ERASE	NO ALC ACTION	ERRATIC SPEED
Q1	Q1	Q1	Open or shorted	X		X	X		X				
Q2	Q2	Q2	Open or shorted	X		X	X		X				
Q3	Q3	Q3	Open or shorted	X		X	X		X				
Q4	Q4	Q4	Open or shorted	X		X	X		X				
Q5	Q5	Q5	Open or shorted	X		X	X		X				
Q6	Q6	Q6	Open or shorted	X		X	X		X				
Q7	Q7	Q7	Open or shorted					X	X			X	
Q8	Q8	Q8	Open or shorted					X	X	X	X		
Q9	Q100	Q100	Open or shorted										X
Q10	Q101	Q101	Open or shorted										X
C1	-	-	Open or shorted				X		X				
-	C1	C31	Open or shorted				X	X	X				
C3	C3	C2	Open or shorted	X	X	X	X	X	X				
C8	C8	C7	Open or shorted	X	X	X	X	X	X				
C9	C9	C8	Open or shorted		X			X					
C11	C11	C10	Open or shorted	X	X	X	X	X	X				
C13	C13	C13	Open or shorted		X			X					
C14	C14	C15	Open or shorted	X		X	X		X				
C15	C15	C16	Open or shorted	X	X	X	X	X	X				
C19	C26	C25	Open or shorted				X	X	X			X	
C20	C23	C23	Open or shorted						X				
C21	C25	C30	Open or shorted									X	
C24	C24	C21	Open or shorted						X	X			
-	-	C24	Open or shorted									X	
C26	C100	C100	Open or shorted										X
D1	D1	D1	Open or shorted									X	
D2	D100	D100	Open or shorted										X
T1	T1	T1	Open or shorted						X	X			
Record/Play head open or shorted				X		X	X		X				
Erase head open or shorted									X	X	X		
Heads bent or misaligned					X	X		X	X		X		
Heads dirty or magnetized					X	X		X	X		X		
Incorrect Record bias (high or low)								X	X		X		
Incorrect erase voltage (high or low)									X	X	X		
Incorrect B+ voltage (high or low)				X	X	X	X	X	X	X	X	X	X
Built-in mike open or shorted							X	X	X				
Record/Play switch defective				X			X			X		X	
Erase head wires reversed (A & B only)									X		X		
Volume control defective				X	X	X							
Defective cassette					X	X		X	X		X		X

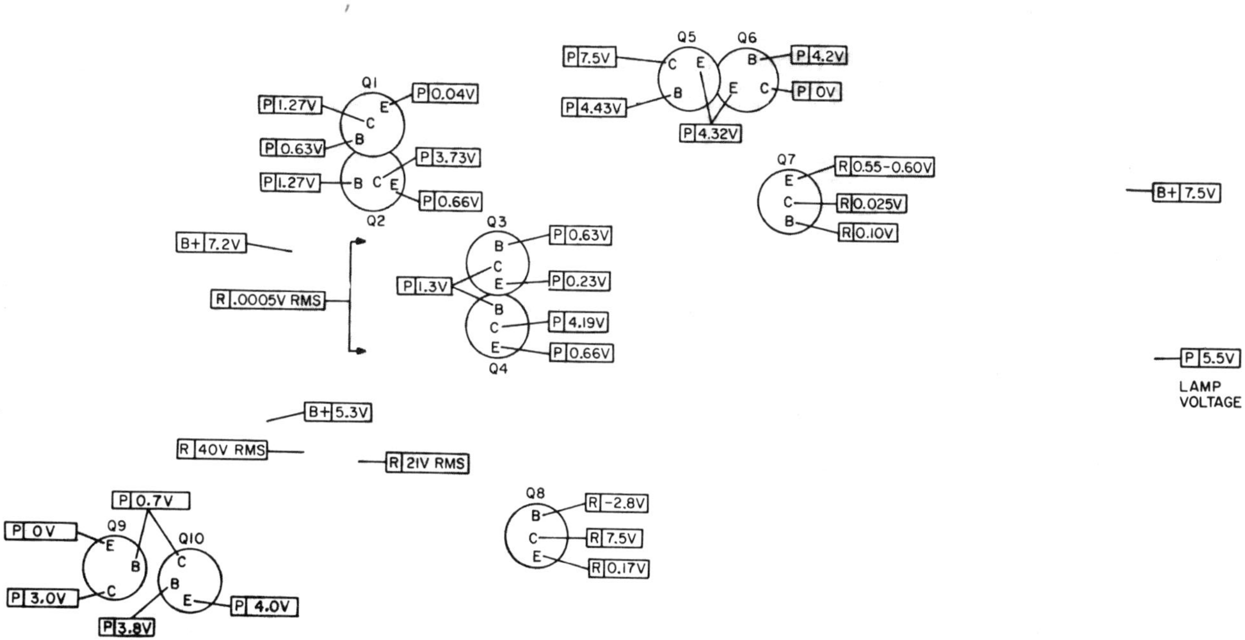

BOTTOM VIEW – CIRCUIT VOLTAGES
MODEL M8700A

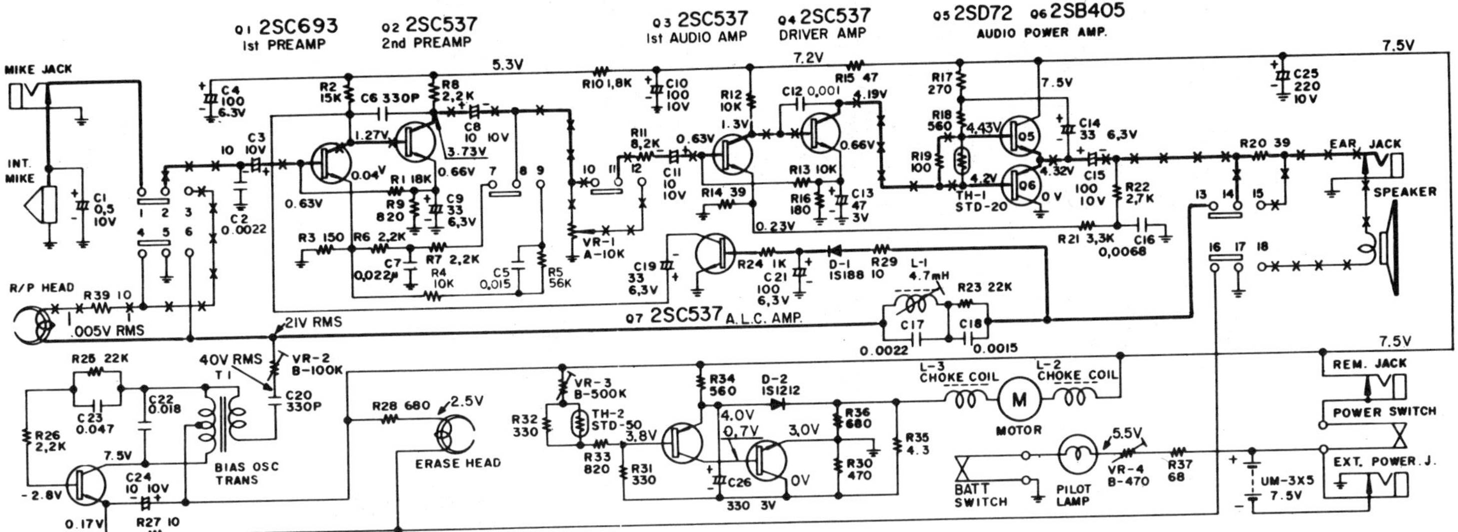

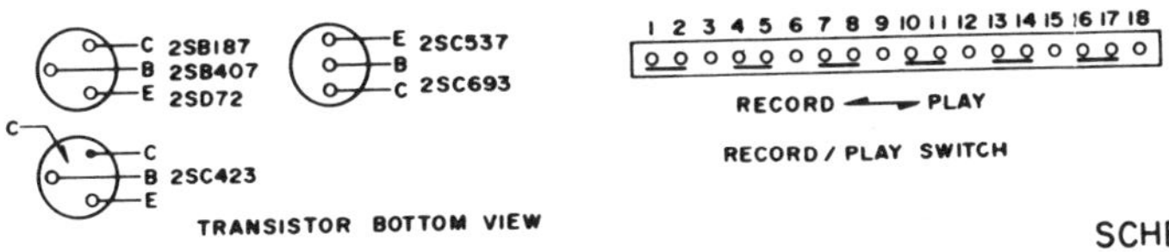

NOTES:

1. (——) RECORD SIGNAL PATH FROM MIKE TO HEAD
2. (-×-×-) PLAYBACK SIGNAL PATH FROM HEAD TO SPEAKER
3. (-✱-✱-) COMMON SIGNAL PATH
4. R/P SWITCH SHOWN IN RECORD MODE
5. VOLTAGE ARE POSITIVE WITH RESPECT TO GROUND TAKEN IN PLAYBACK MODE EXCEPT FOR Q7 & Q8 AND BIAS VOLTAGES WHICH ARE TAKEN IN RECORD MODE. RMS VOLTAGES TAKEN WITH AC VTVM.

SCHEMATIC DIAGRAM
MODEL M8700A

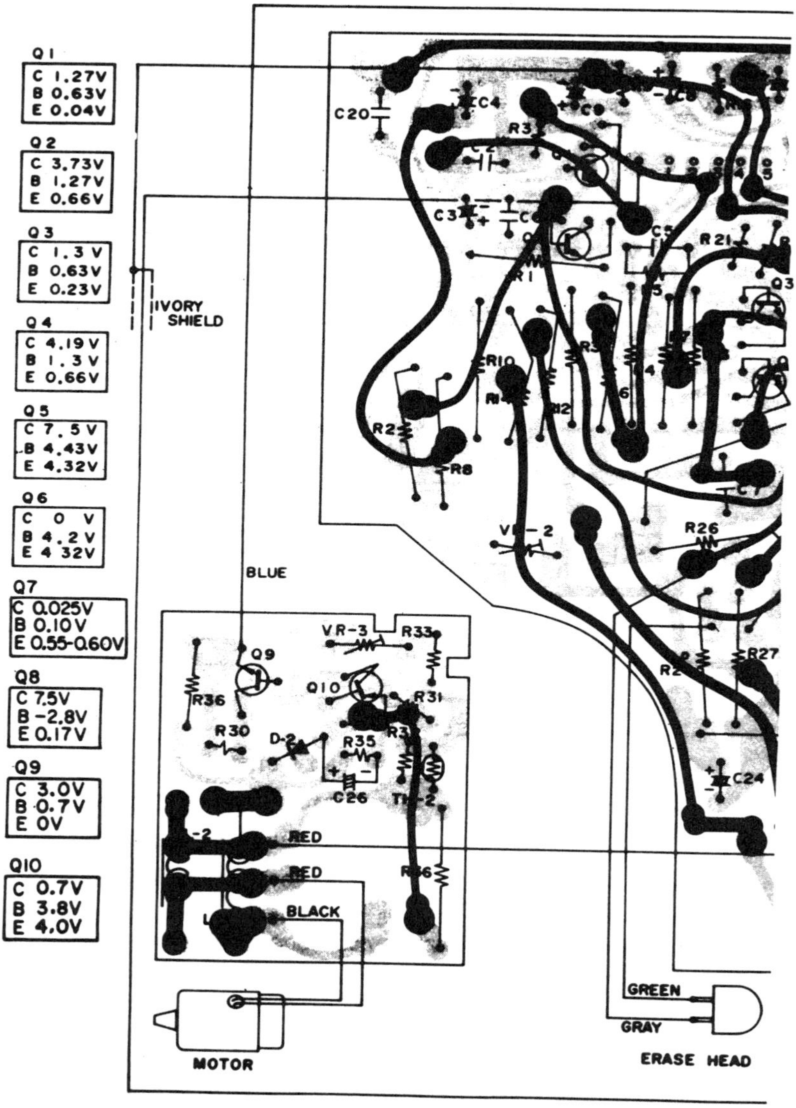

WIRING DIAGRAM—BOTTOM VIEW
MODEL M8700A

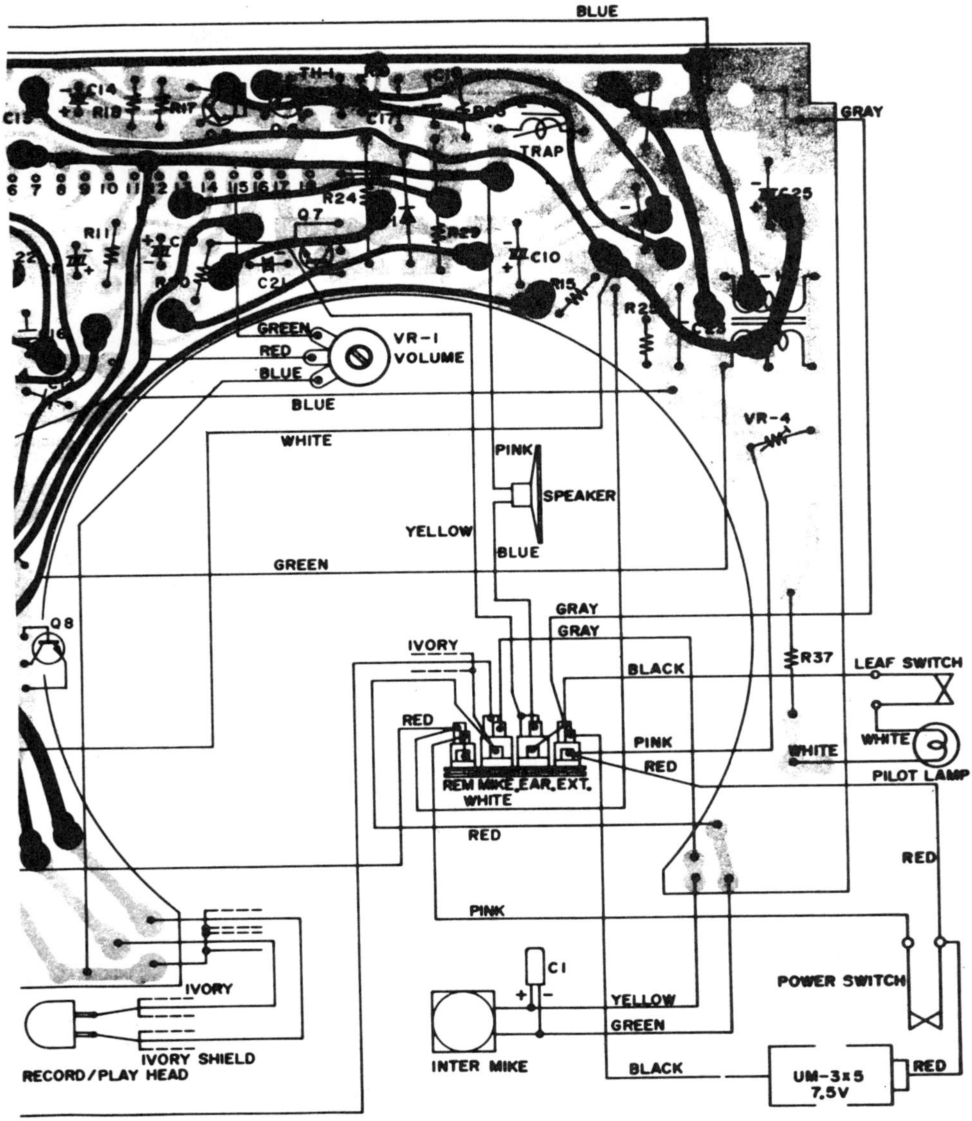
BLUE
GRAY
TRAP
R24
Q7
R11
C10
C21
R15
C25
VR-1
VOLUME
GREEN
RED
BLUE
BLUE
WHITE
VR-4
PINK
SPEAKER
YELLOW
BLUE
GREEN
Q8
GRAY
GRAY
IVORY
R37
BLACK
LEAF SWITCH
RED
WHITE
PINK
WHITE
RED
PILOT LAMP
REM MIKE EAR EXT.
WHITE
RED
RED
PINK
C1
IVORY
POWER SWITCH
YELLOW
GREEN
IVORY SHIELD
RECORD/PLAY HEAD
INTER MIKE
BLACK
UM-3x5
7.5V
RED

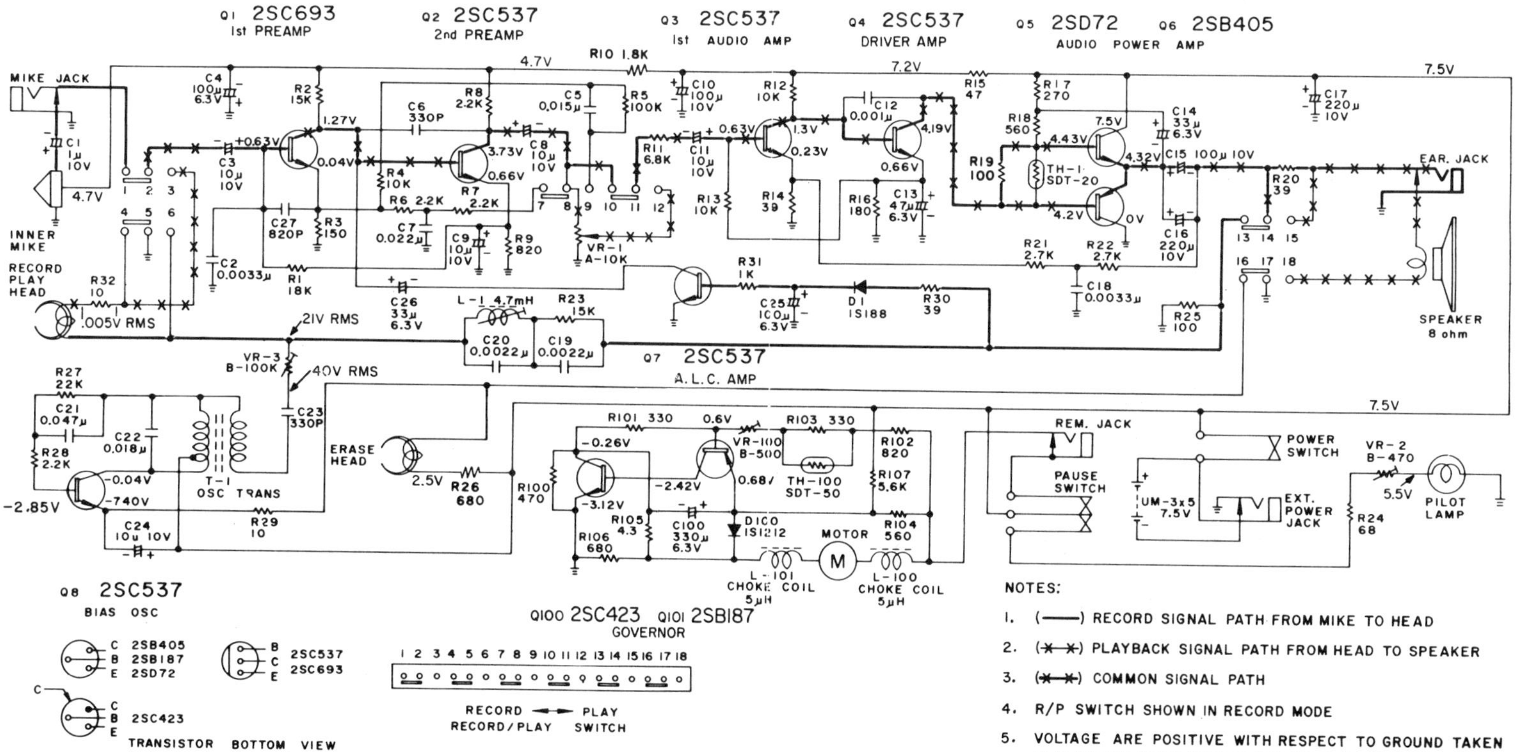

NOTES:

1. (——) RECORD SIGNAL PATH FROM MIKE TO HEAD
2. (✕—✕) PLAYBACK SIGNAL PATH FROM HEAD TO SPEAKER
3. (✕━✕) COMMON SIGNAL PATH
4. R/P SWITCH SHOWN IN RECORD MODE
5. VOLTAGE ARE POSITIVE WITH RESPECT TO GROUND TAKEN IN PLAYBACK MODE EXCEPT FOR Q7 & Q8 AND BIAS VOLTAGES WHICH ARE TAKEN IN RECORD MODE. RMS VOLTAGES TAKEN WITH AC VTVM.

SCHEMATIC DIAGRAM
MODEL M8700B

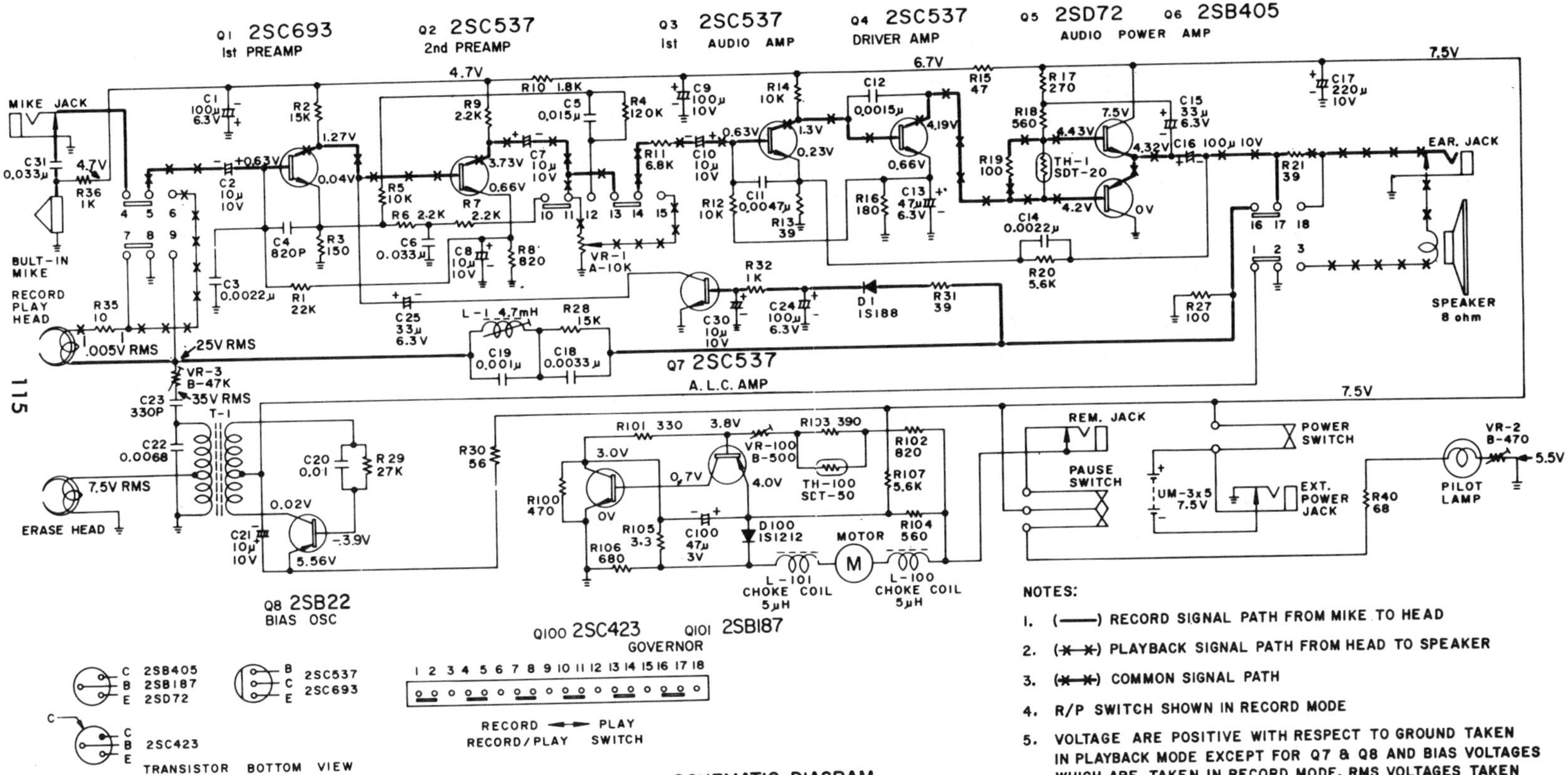

SCHEMATIC DIAGRAM
MODEL M8700C

NOTES:

1. (——) RECORD SIGNAL PATH FROM MIKE TO HEAD
2. (—×—×) PLAYBACK SIGNAL PATH FROM HEAD TO SPEAKER
3. (—×—×) COMMON SIGNAL PATH
4. R/P SWITCH SHOWN IN RECORD MODE
5. VOLTAGE ARE POSITIVE WITH RESPECT TO GROUND TAKEN IN PLAYBACK MODE EXCEPT FOR Q7 & Q8 AND BIAS VOLTAGES WHICH ARE TAKEN IN RECORD MODE. RMS VOLTAGES TAKEN WITH AC VTVM.

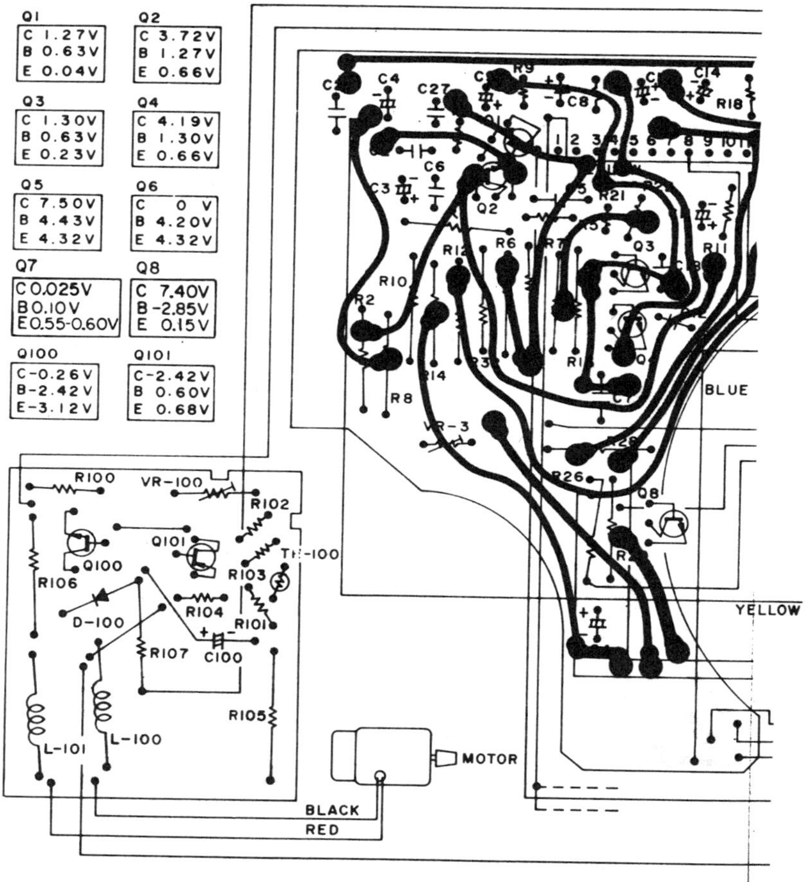

WIRING DIAGRAM — BOTTOM VIEW
MODEL M8700B

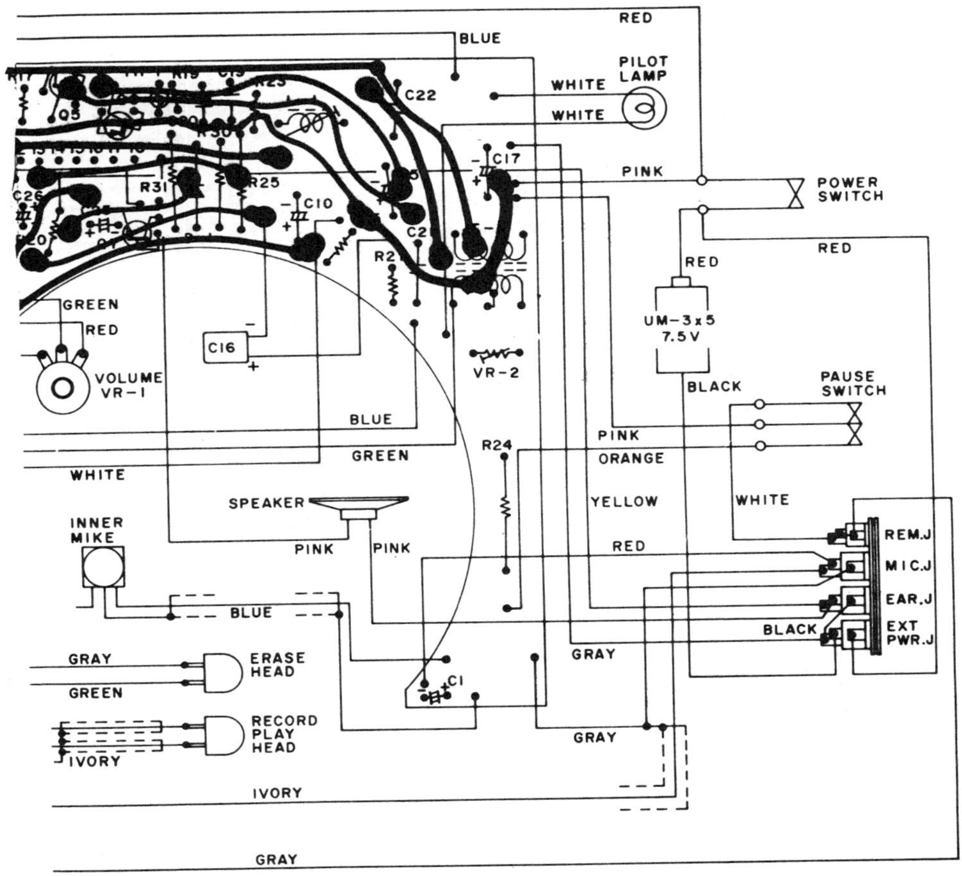
RED
BLUE
PILOT LAMP
WHITE
WHITE
C22
Q5
R30
C17
PINK
POWER SWITCH
R31
R25
C26
C10
R20
RED
GREEN
RED
C16
VOLUME VR-1
UM-3x5 7.5V
VR-2
BLACK
PAUSE SWITCH
BLUE
PINK
R24
GREEN
ORANGE
WHITE
SPEAKER
YELLOW
WHITE
INNER MIKE
PINK
PINK
RED
REM.J
MIC.J
EAR.J
BLUE
BLACK
EXT PWR.J
GRAY
ERASE HEAD
GRAY
GREEN
C1
RECORD PLAY HEAD
GRAY
IVORY
IVORY
GRAY

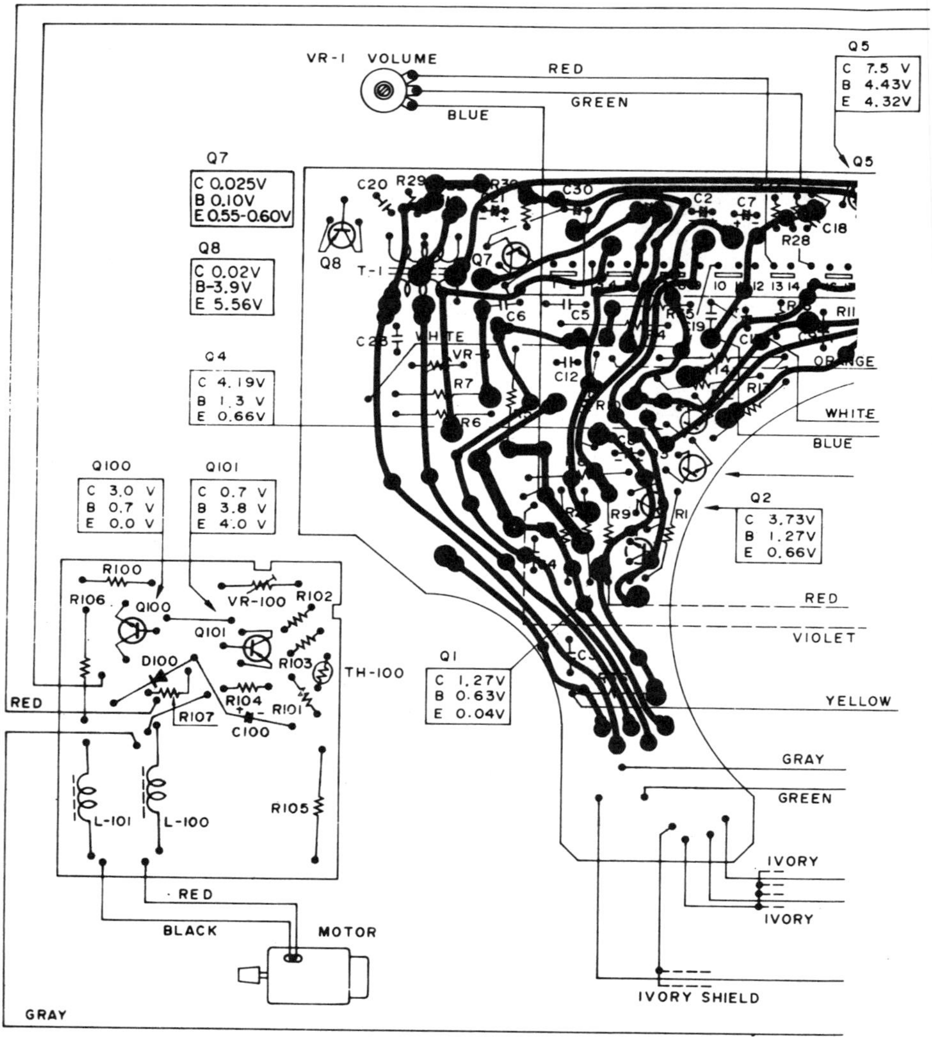
VR-1 VOLUME
RED
GREEN
BLUE
Q5
C 7.5 V
B 4.43V
E 4.32V
Q7
C 0.025V
B 0.10V
E 0.55-0.60V
Q8
C 0.02V
B-3.9V
E 5.56V
Q4
C 4.19V
B 1.3 V
E 0.66V
Q100
C 3.0 V
B 0.7 V
E 0.0 V
Q101
C 0.7 V
B 3.8 V
E 4.0 V
Q2
C 3.73V
B 1.27V
E 0.66V
Q1
C 1.27V
B 0.63V
E 0.04V
C20
R29
C30
C2
C7
C18
R28
T-1
C6
C5
C19
R11
C28
WHITE
VR-3
R7
R6
C12
R9
R1
ORANGE
WHITE
BLUE
RED
VIOLET
YELLOW
GRAY
GREEN
IVORY
IVORY
IVORY SHIELD
R100
R106
Q100
VR-100
R102
Q101
D100
R103
TH-100
R104
R101
R107
C100
RED
L-101
L-100
R105
RED
BLACK
MOTOR
GRAY

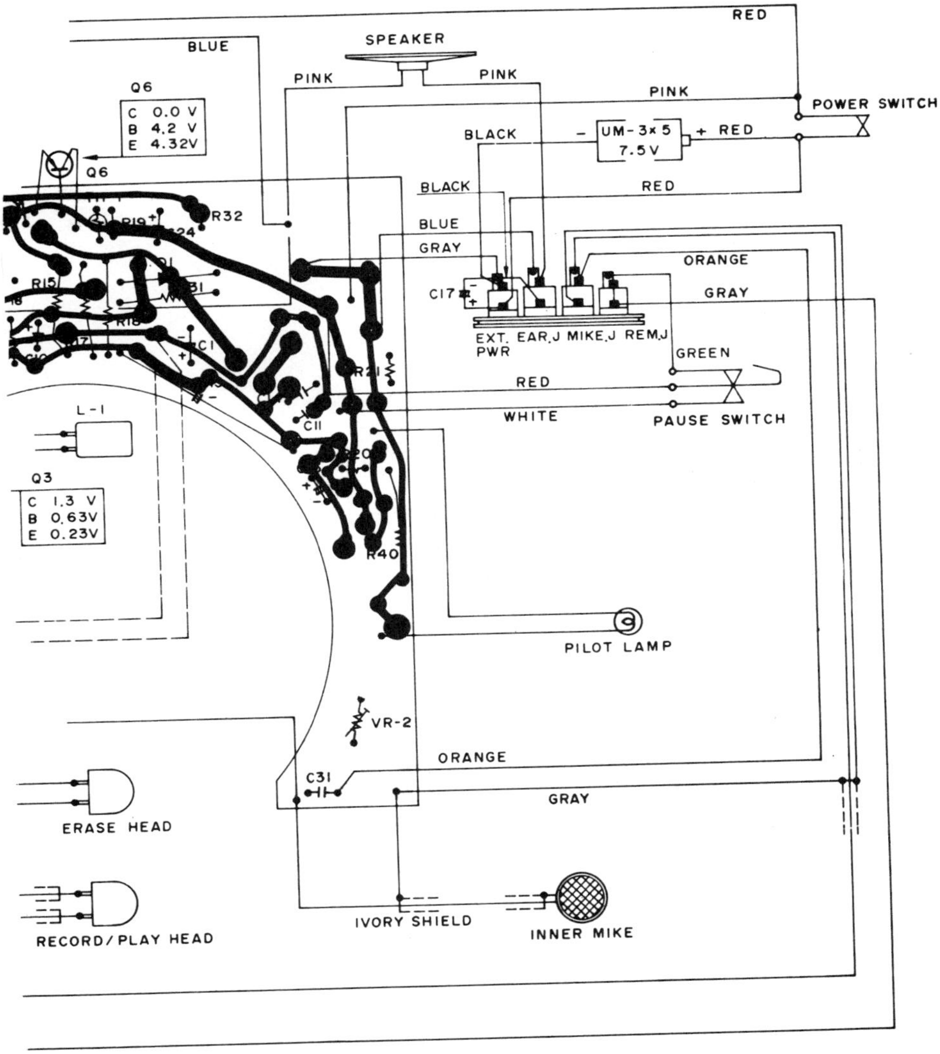
RED
BLUE
SPEAKER
PINK
PINK
PINK
POWER SWITCH
Q6
C 0.0 V
B 4.2 V
E 4.32V
BLACK
UM-3x5
7.5V
RED
Q6
BLACK
RED
R32
R19
BLUE
GRAY
ORANGE
R15
C17
GRAY
R18
C1
EXT. PWR
EAR.J
MIKE.J
REM.J
GREEN
R21
RED
WHITE
PAUSE SWITCH
L-1
C11
Q3
C 1.3 V
B 0.63V
E 0.23V
R40
PILOT LAMP
VR-2
ORANGE
C31
GRAY
ERASE HEAD
IVORY SHIELD
INNER MIKE
RECORD/PLAY HEAD

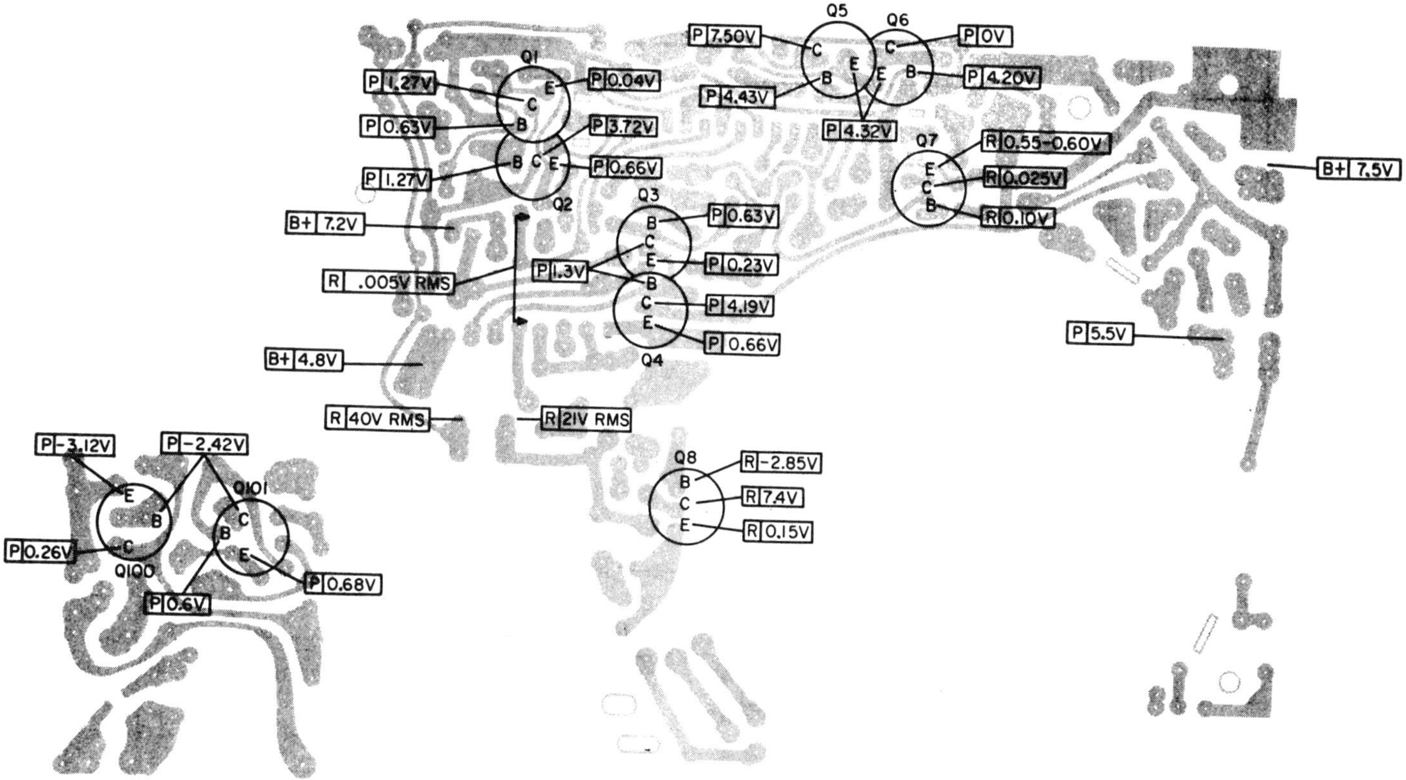

BOTTOM VIEW—CIRCUIT VOLTAGES
MODEL M8700B

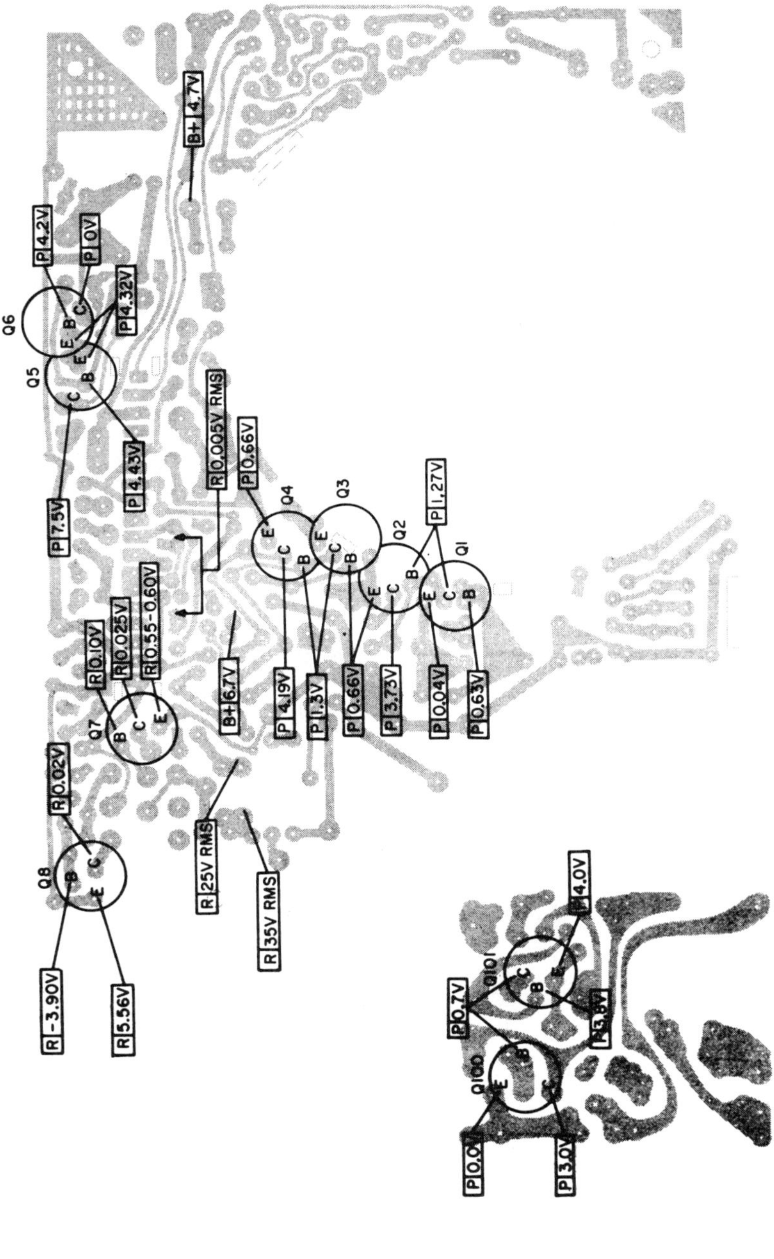

B+ 4.7V
P 4.2V
P 0V
P 4.32V
Q6
Q5
P 4.43V
P 7.5V
R 0.005V RMS
P 0.66V
Q4
Q3
Q2
Q1
P 1.27V
R 0.10V
R 0.025V
R 0.55-0.60V
Q7
B+ 6.7V
P 4.19V
P 1.3V
P 0.66V
P 3.73V
P 0.04V
P 0.63V
R 0.02V
Q8
R 25V RMS
R 35V RMS
R -3.90V
R 5.56V
P 4.0V
Q101
P 0.7V
Q100
P 0.0V
P 3.0V

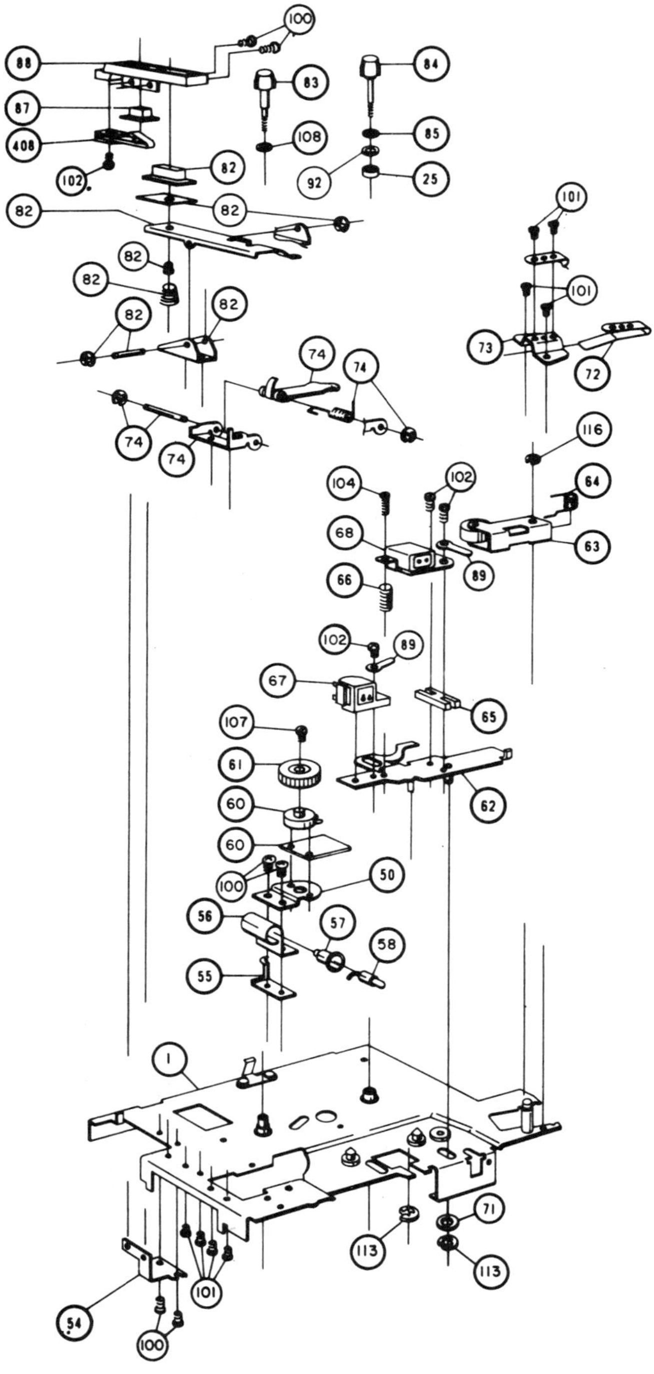

100
88
83
84
87
108
85
408
92
25
102
82
82
82
82
82
82
82
101
101
73
72
74
74
74
74
116
104
102
64
68
63
89
66
102
89
67
65
107
61
62
60
60
100
50
56
57
58
55
1
71
113
113
101
54
100

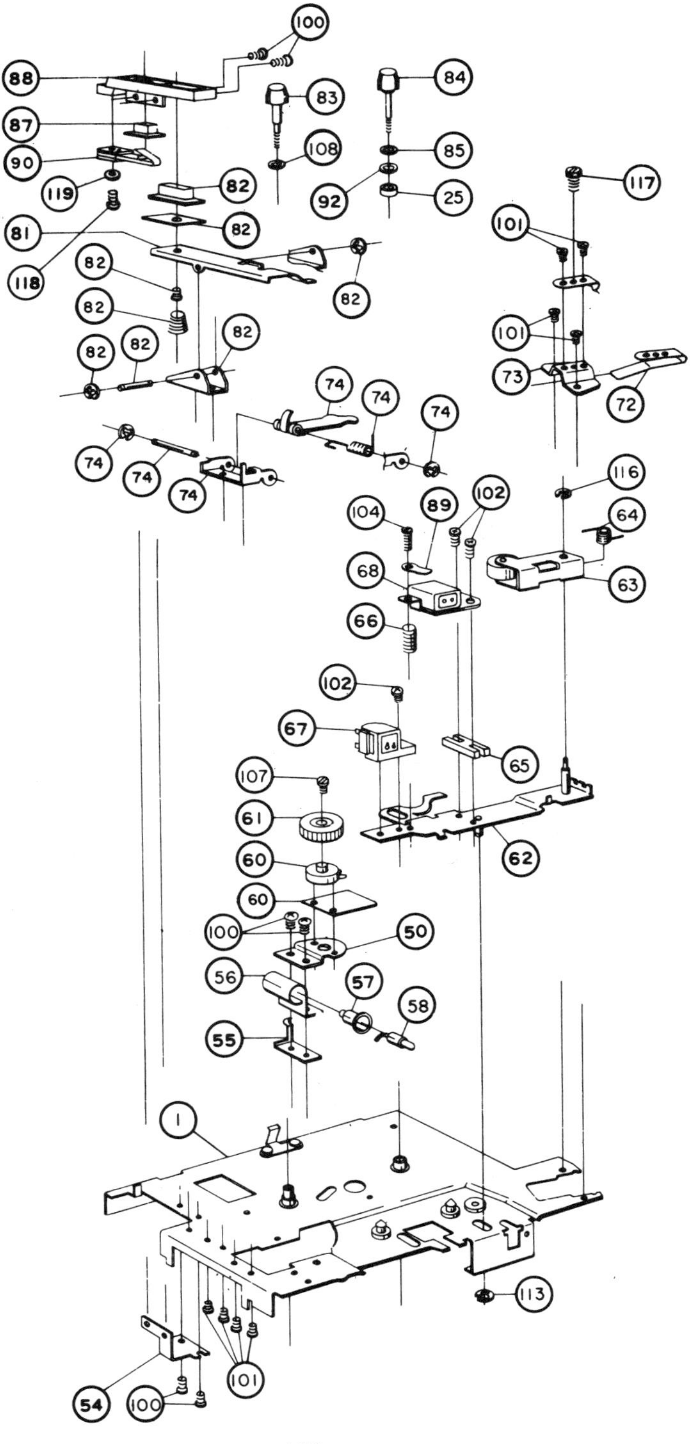
100
88
84
83
87
108
85
90
92
25
119
82
82
117
81
118
101
82
82
82
101
82
82
82
73
72
74
74
74
74
74
74
116
104
89
102
64
68
63
66
102
67
65
107
61
62
60
60
100
50
56
57
58
55
1
113
101
54
100

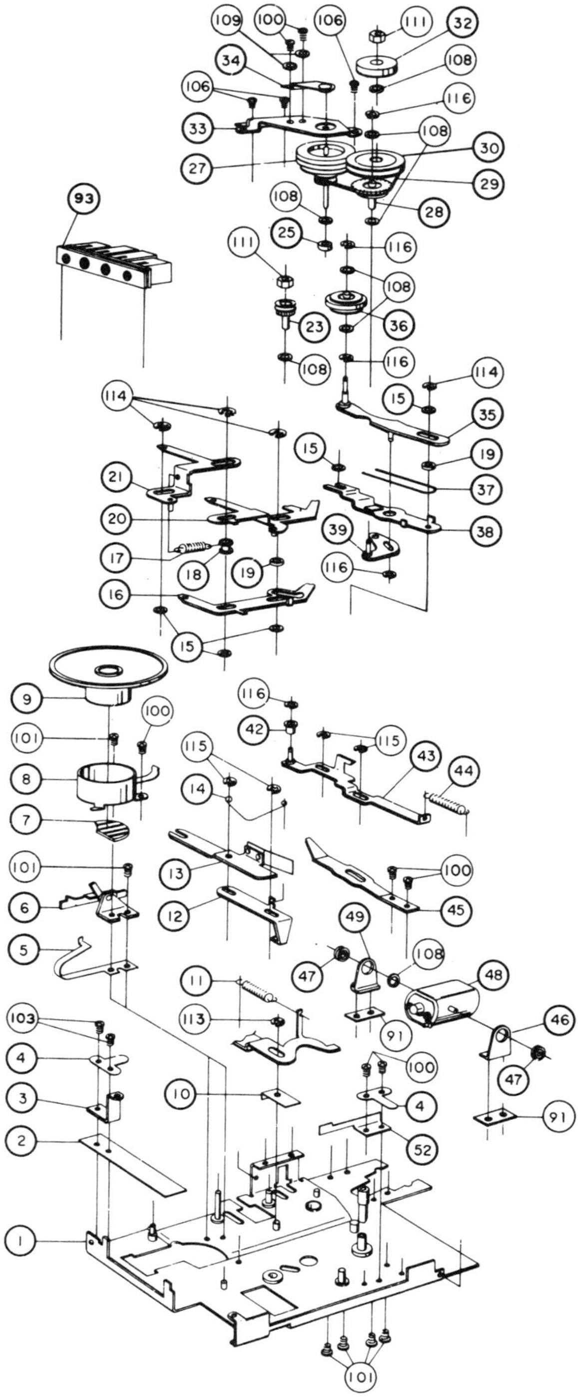

109
100
106
111
32
34
108
106
116
33
108
30
27
29
93
108
28
25
111
116
108
23
36
108
116
114
114
15
35
15
19
21
37
20
38
39
17
18
19
116
16
15
9
116
100
42
101
115
115
43
44
8
14
7
13
101
100
6
45
12
49
5
108
47
48
11
113
91
46
103
4
100
47
3
10
4
91
2
52
1
101

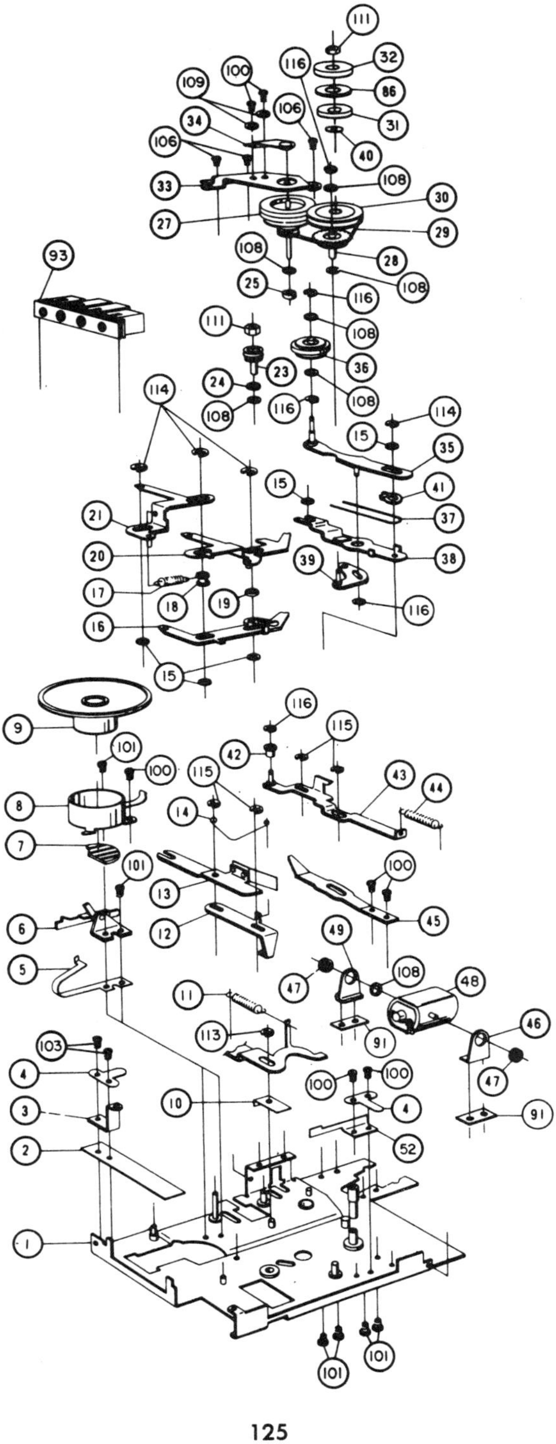
111
32
116
100
86
109
106
31
34
40
106
108
33
30
27
29
93
108
28
25
108
116
111
108
23
36
24
108
116
108
114
114
15
35
15
41
37
21
20
38
39
17
18
19
116
16
15
116
9
115
101
42
100
115
43
44
8
14
7
101
100
13
45
6
12
49
5
108
48
47
11
91
46
113
103
4
100
100
47
10
3
4
91
2
52
1
101
101

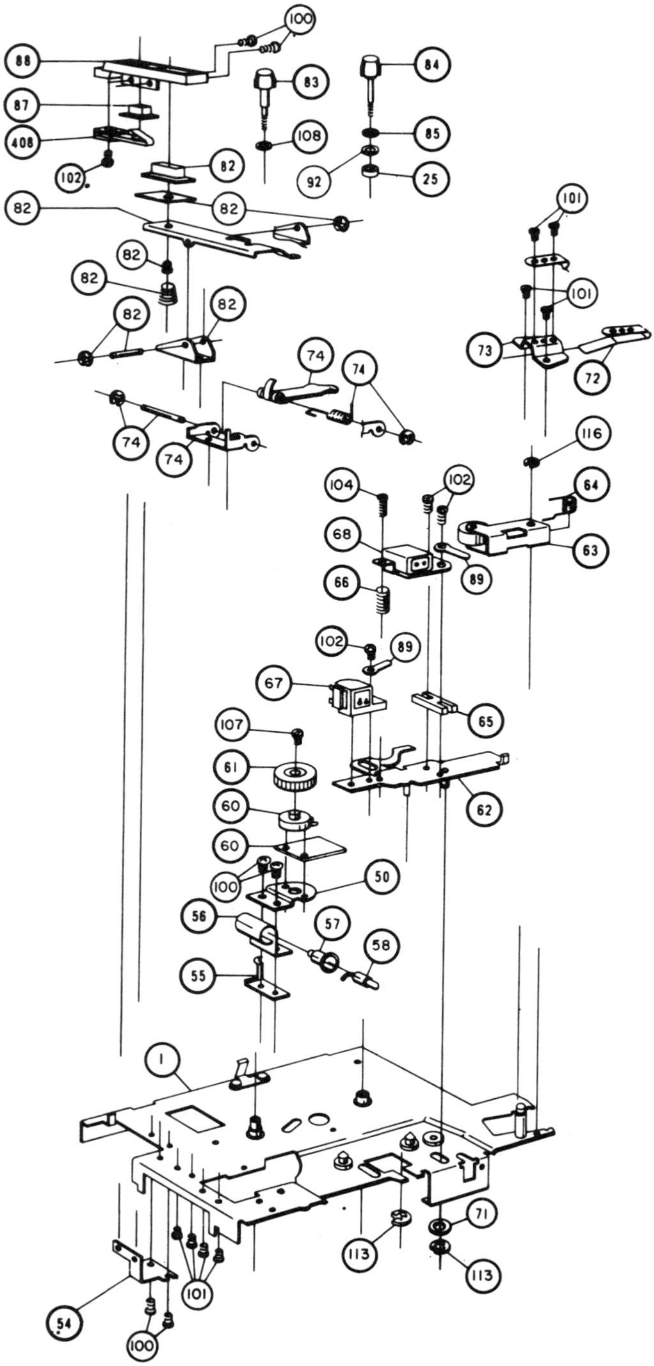

EXPLODED VIEW - PARTS ABOVE DECK
MODEL M8700C

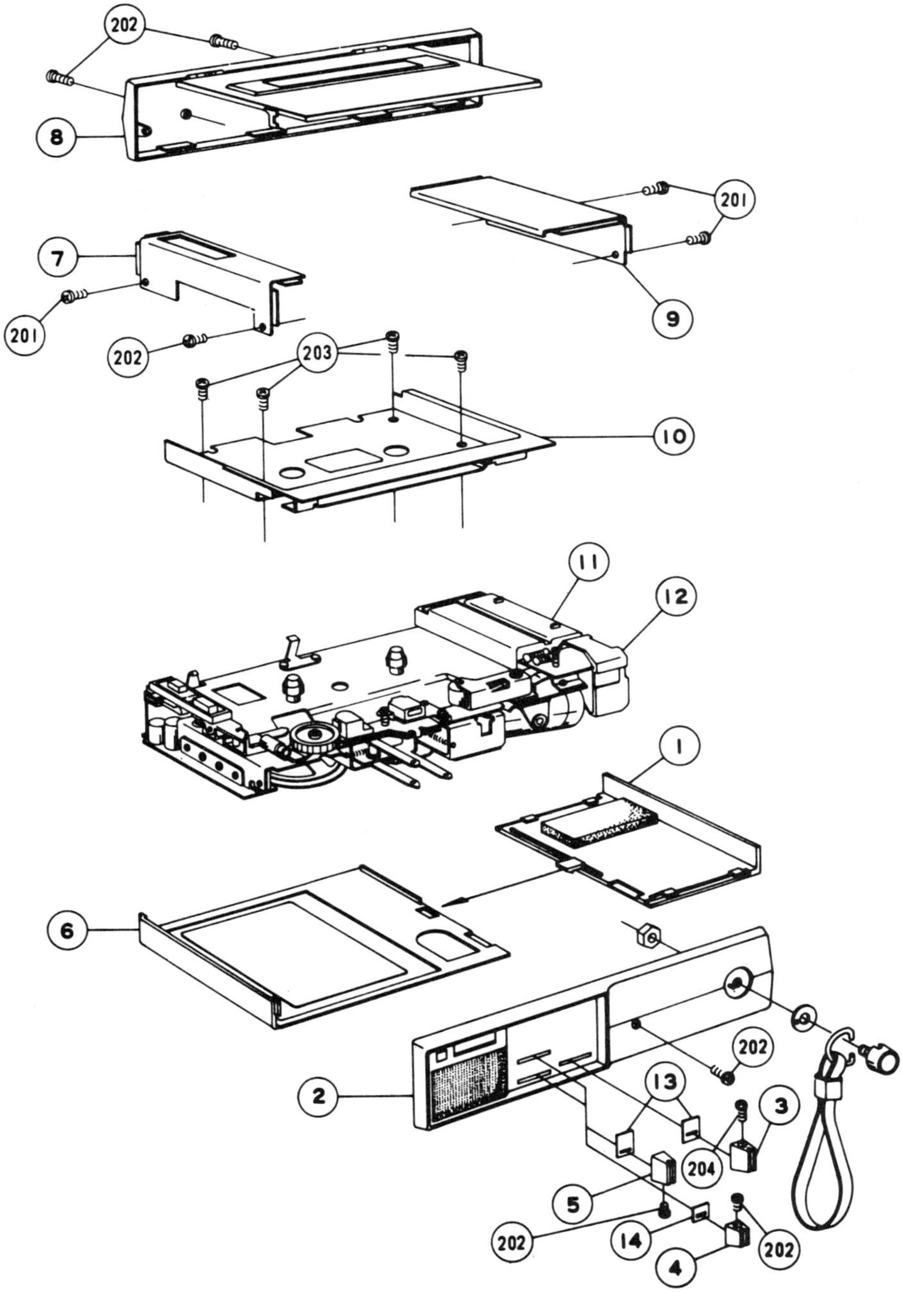
202
8
201
7
201
202
203
9
10
11
12
1
6
2
202
13
3
204
5
202
14
202
4

REPLACEMENT PARTS LIST

CAT. NO.	REF. NO.	DESCRIPTION	VERSION A	B	C
		MECHANICAL PARTS			
RT9856	1	Chassis Assy	X	X	
*EA92X3	1	Chassis Assy			X
RT6500	2	Jack Insulator	X	X	X
*EA2X91	3	Circuit Board Mtg. Brkt.	X	X	X
*EA3X49	4	Wire Clamp	X	X	X
RT6501	5	Record Spring	X	X	X
RT6502	6	Record Lever Assy	X	X	X
RT6503	7	Speaker Cushion	X	X	X
RT6504	8	Speaker Brkt	X	X	X
RB2908	9	Speaker	X	X	X
RT6506	10	Slide Base Spring	X	X	X
RT6506	11	Slide Base Spring	X	X	X
RT6507	12	Interlock Lever	X	X	X
RT6508	13	Record SE Lever	X	X	X
RT6509	14	Interlock Lever Spring	X	X	X
RT6510	15	Washer	X	X	X
RT6511	16	Play Lever	X	X	X
RT6512	17	Record Lever Spring	X	X	X
RT6513	18	Record Lever Spacer	X	X	X
RT6514	19	FWD Lever Spacer	X	X	X
RT6515	20	FWD Lever Assy	X	X	X
RT6516	21	Record Lever Assy	X	X	X
RT6517	23	RWD Pulley	X	X	
*EA6X91	23	RWD Pulley			X
*EA1X76	24	Slide Base Pin Washer			X
RT6519	25	Ball Bearing	X	X	X
RT6520	27	Flywheel	X	X	X
RT6521	28	RWD Idler	X	X	
*EA1X78	28	RWD Idler			X
RT6522	29	Drive Belt	X	X	X
RT6523	30	Take Up Clutch (Lower)	X	X	
*EA3X50	30	Take Up Clutch (Lower)			X
*EA3X52	31	Clutch Felt	X	X	X
RT6524	32	Take Up Clutch (Upper)	X	X	X
RT6525	33	Capstan Bearing Plate Assy	X	X	X
RT6526	34	Capstan Thrust Plate	X	X	X
RT6527	35	FWD Idler Arm Assy	X	X	X
RT6528	36	FWD Idler Roller	X	X	X
RT6529	37	RWD Spring	X	X	X
RT6531	39	RWD Idler Arm Assy	X	X	X
*EA1X77	40	Clutch Spindle Washer		X	X
*EA2X92	41	Grip Ring (Horseshoe)			X
RT6533	42	Roller Pin	X	X	X
RT6535	44	Roller Pin Arm Spring	X	X	X
RT6536	45	Take Up Clutch Spring	X	X	X
RT6537	46	Motor Bracket (L)	X	X	X
RT6538	47	Motor Shaft Grommet	X	X	X
RT6540	48	Motor	X	X	
*EA72X3	48	Motor			X
RT6539	49	Motor Bracket (R)	X	X	X
RT6541	50	Volume Control Bracket	X	X	X
RT6542	52	FWD Idler Arm Spring	X	X	X
*EA2X93	54	Button Mtg. Bracket	X	X	X
RT6543	55	Cassette Door Lock Spring	X	X	X
RT6544	56	Lamp Bracket	X	X	X
RT6546	57	Lamp Holder - Rubber -	X	X	X
RT6545	58	Pilot Lamp	X	X	X
RT6548	61	Volume Knob	X	X	X
RT6549	62	Slide Base Assy	X	X	
*EA5X4	62	Slide Base Assy			X
RT6550	63	Pinchwheel Assy	X	X	
*EA5X5	63	Pinchwheel Assy			X
RT6551	64	Pinchwheel Spring	X	X	
*EA3X51	64	Pinchwheel Spring			X

CAT. NO.	REF. NO.	DESCRIPTION	VERSION A	B	C
		MECHANICAL PARTS (CONT'D)			
RT6552	65	Head Spacer	X		
RT8504	65	Head Spacer		X	X
RT6553	66	Azimuth Adjust Spring	X		
RT8505	66	Azimuth Adjust Spring		X	X
RT6554	67	Erase Head	X	X	
*EA62X38	67	Erase Head			X
RT6555	68	Record/Play Head	X		
RT8506	68	Record/Play Head		X	
*EA62X39	68	Record/Play Head			X
RT6556	69	Head Spacer	X		
*EA1X74	71	Washer, Slide Base Mtg.			X
RT6557	72	Motor Tension Spring	X	X	X
RT6558	73	Motor Spring Bracket	X	X	X
RT6560	74	Cassette Eject Assy	X	X	X
RT8508	82	Eject Lever Assy	X	X	X
RT6569	83	Supply Hub Assy	X	X	X
RT6570	84	Take Up Hub Assy	X	X	X
RT6571	85	Hub Spacer	X	X	X
RT6572	86	Washer		X	X
RT6573	87	Battery Pushbutton	X		
RT8510	87	Pause Pushbutton		X	X
RT6574	88	Button Bracket	X		
RT8511	88	Button Bracket		X	X
RT6575	90	Leaf Switch	X		
RT8512	90	Leaf Switch		X	X
RT8507	92	Bearing Cushion	X	X	X
		MISCELLANEOUS HARDWARE			
RT7948	100	Screw 2x1.8mm	X	X	X
RT7948	101	Screw 2x3mm	X	X	X
RT7948	102	Screw 2x4mm	X	X	X
RT7948	104	Azimuth Adjust Screw 2x10mm	X	X	X
RT7948	105	Screw 2x6mm	X		
RT7948	106	Screw 2x4mm	X	X	X
RT7948	107	Volume Knob Screw 1.7x4mm	X	X	X
RT7949	108	Washer 2.1x4x13mm	X	X	X
RT7949	109	Washer 2mm	X	X	X
RT7949	111	Nut 1.7mm	X	X	X
RT7949	113	"E" Ring 3mm	X	X	X
RT6288	114	"E" Ring 2.5mm	X	X	X
RT6287	115	"E" Ring 2mm	X	X	X
RT7949	116	"E" Ring 1.5mm	X	X	X
RT7948	117	Screw, Motor Control	X	X	X
EA1X75	201	Screw, End Panel 2x1.8mm	X	X	X
EA1X75	202	Screw, End Panel 2x4mm	X	X	X
EA1X75	203	Screw, Chassis Panel 2x1.8mm	X	X	X
EA1X75	204	Screw, Control Knob 2x8mm	X	X	X

REPLACEMENT PARTS LIST (cont'd)

CAPACITORS

CAT. NO.	REF. NO.	DESCRIPTION		VERSION A	B	C
RT2854	C12	.5mfd/10V	(A)	X		
RT8353	C1	1mfd/10V	(A)		X	
RT4769	C1	100mfd/6.3	(E)			X
RT2309	C2	.0033mfd/50V	(M)	X	X	
RT2306	C2	10mfd/10V	(E)			X
RT2306	C3	10mfd/10V	(E)	X	X	
RT4570	C3	.0022mfd/50V	(M)			X
RT4769	C4	100mfd/6.3V	(E)	X	X	
RT1980	C4	820pf/50V	(C)			X
RT5439	C5	.015mfd/50V	(M)	X	X	X
RT1981	C6	330pf/50V	(M)	X	X	
RT4562	C6	.033mfd/50V	(M)			X
RT2310	C7	.022mfd/50V	(M)	X	X	
RT2306	C7	10mfd/10V	(E)			X
RT2306	C8	10mfd/10V	(E)	X	X	X
RT4560	C9	33mfd/6.3V	(E)	X		
RT2306	C9	10mfd/10V	(E)		X	
RT5444	C9	100mfd/10V	(E)			X
RT5444	C10	100mfd/10V	(E)	X	X	
RT2306	C10	10mfd/10V	(E)			X
RT2306	C11	10mfd/10V	(E)	X	X	
RT2860	C11	.0047mfd/50V	(M)			X
RT1980	C12	820pf/50V	(C)	X		
RT4773	C12	.001mf/50V	(M)		X	
RT5437	C12	.0015mf/50V	(M)			X
RT5775	C13	47mfd/3V	(E)	X		
RT4632	C13	47mfd/6.3	(E)		X	X
RT4560	C14	33mfd/6.3	(E)	X	X	
RT4570	C14	.0022mfd/50V	(M)			X
RT5444	C15	100mfd/10V	(E)	X	X	
RT4560	C15	33mfd/6.3V	(E)			X
RT5438	C16	.0068mfd/50V	(M)	X		
RT2313	C16	220mfd/10V	(E)		X	
RT5444	C16	100mfd/10V	(E)			X
RT4570	C17	.0022mfd/50V	(M)	X		
RT2313	C17	220mfd/10V	(E)		X	X
RT5437	C18	.0015mfd/50V	(M)	X		
RT2309	C18	.0033mfd/50V	(M)		X	X
RT4560	C19	33mfd/6.3V	(E)	X		
RT4570	C19	.0022mfd/50V	(M)		X	
RT4773	C19	.001mfd/50V	(M)			X
EA2505	C20	330pf/50V	(M)	X		
RT4570	C20	.0022mfd/50V	(M)		X	
RT3177	C20	.01mfd/50V	(M)			X
RT5776	C21	100mfd/6.3V	(E)	X		
RT2308	C21	.047mfd/50V	(M)		X	
RT2306	C21	10mfd/10V	(E)			X
RT6156	C22	.018mfd/50V	(M)	X	X	
RT5438	C22	.0068mfd/150V	(M)			X
RT2308	C23	.047mfd/50V	(M)	X		
EA2505	C23	330pf/50V	(M)		X	
RT1981	C23	330pf/50V	(C)			X
RT2306	C24	10mfd/10V	(E)	X	X	
RT5776	C24	100mfd/6.3V	(E)			X
RT2313	C25	220mfd/10V	(E)	X		
RT5776	C25	100mfd/6.3V	(E)		X	
RT4560	C25	33mfd/6.3V	(E)			X
RT5777	C26	330mfd/3V	(E)	X		
RT4560	C26	33mfd/6.3V	(E)		X	
RT2306	C30	10mfd/10V	(E)			X
RT4562	C31	.033mfd/50V	(M)			X

RESISTORS

CAT. NO.	REF. NO.	DESCRIPTION	VERSION A	B	C
*EA13X11	R1	18K	X	X	
EA13X11	R1	22K			X
EA13X11	R2	15K	X	X	X
EA13X11	R3	150 ohms	X	X	X
EA13X11	R4	10K	X	X	
EA13X11	R4	120K			X
EA13X11	R5	100K		X	
EA13X11	R5	10K			X
EA13X11	R6, 7	2.2K	X	X	X
EA13X11	R8	2.2K	X	X	
EA13X11	R8	820 ohms			X
EA13X11	R9	820 ohms	X	X	
EA13X11	R9	2.2K			X
EA13X11	R10	1.8K	X	X	X
EA13X11	R11	8.2K	X		
EA13X11	R11	6.8K		X	X
EA13X11	R12	10K	X	X	X
EA13X11	R13	10K	X	X	
EA13X11	R13	39 ohms			X
EA13X11	R14	39 ohms	X	X	
EA13X11	R14	10K			X
EA13X11	R15	47 ohms	X	X	X
EA13X11	R16	180 ohms	X		X
EA13X11	R16	180 ohms		X	
EA13X11	R17	270 ohms	X	X	X
EA13X11	R18	560 ohms	X	X	X
EA13X11	R19	100 ohms	X	X	X
EA13X11	R20	39 ohms	X	X	
EA13X11	R20	5.6K			X
EA13X11	R21	3.3K	X		
EA13X11	R21	2.7K		X	
EA13X11	R21	39 ohms			X
EA13X11	R22	2.7K	X	X	
EA13X11	R23	22K	X		
EA13X11	R23	15K		X	
EA13X11	R24	1K	X		
EA13X11	R24	68 ohms		X	
EA13X11	R25	22K	X		
EA13X11	R25	100 ohms		X	
EA13X11	R26	2.2K	X		
EA13X11	R26	680 ohms		X	
EA13X11	R27	10 ohms	X		
EA13X11	R27	22K		X	
EA13X11	R27	100 ohms			X
EA13X11	R28	680 ohms	X		
EA13X11	R28	2.2K		X	
EA13X11	R28	15K			X
EA13X11	R29	10 ohms	X		
EA13X11	R29	10 ohms		X	
EA13X11	R29	27K			X
EA13X11	R30	470 ohms	X		
EA13X11	R30	39 ohms		X	
EA13X11	R30	56 ohms			X
EA13X11	R31	330 ohms	X		
EA13X11	R31	1K		X	
EA13X11	R31	39 ohms			X
EA13X11	R32	330 ohms	X		
EA13X11	R32	10 ohms		X	
EA13X11	R32	1K			X
EA13X11	R33	820 ohms	X		
EA13X11	R34	560 ohms	X		

REPLACEMENT PARTS LIST (cont'd)

CAT. NO.	REF. NO.	DESCRIPTION	VERSION A	B	C
		RESISTORS (CONT'D)			
*EA13X11	R35	4.3 ohms	X		
EA13X11	R35	10 ohms			X
EA13X11	R36	680 ohms	X		
EA13X11	R36	1K			X
EA13X11	R37	68 ohms	X		
EA13X11	R38	10 ohms	X		
EA13X11	R40	68 ohms			X
EA13X11	R100	470 ohms		X	X
EA13X11	R101	330 ohms		X	X
EA13X11	R102	820 ohms		X	X
EA13X11	R103	330 ohms		X	
EA13X11	R103	390 ohms			X
EA13X11	R104	560 ohms		X	X
EA13X11	R105	4.3 ohms		X	
EA13X11	R105	3.3 ohms			X
EA13X11	R106	680 ohms		X	X
EA13X11	R107	5.6K		X	
EA13X11	R107	5.6K			X
RT8518	VR-1	Volume Control 10K	X	X	X
RT6583	VR-2	Bias Adjust Pot 100K	X		
RT6582	VR-2	Lamp Adjust Pot 470 ohms		X	X
RT6565	VR-3	Speed Adjust Pot 500K	X		
RT6583	VR-3	Bias Adjust Pot 100K		X	
*EA49X176	VR-3	Bias Adjust Pot 47K			X
RT6582	VR-4	Lamp Adjust Pot 470 ohms	X		
RT8514	VR-100	Speed Adjust Pot 500 ohms		X	X
		TRANSISTORS & DIODES			
RT5435	Q1	1st Audio Preamp, 2SC693	X	X	X
RT7943	Q2	2nd Audio Preamp, 2SC537	X	X	X
RT7943	Q3	1st Audio Amp, 2SC537	X	X	X
RT7943	Q4	Audio Driver, 2SC537	X	X	X
RT7944	Q5	Power Amp, 2SD72	X	X	X
RT4625	Q6	Power Amp, 2SB405	X	X	X
RT7943	Q7	ALC Amp, 2SC537	X	X	X
RT7943	Q8	Bias Osc., 2SC537	X	X	
RT2331	Q8	Bias Osc., 2SB22			X
RT7945	Q9, Q100	Governor Stage, 2SC423	X	X	X
RT3468	Q10, Q101	Governor Stage, 2SB187	X	X	X
RT6199	TH-1	Thermistor, SDT-20	X	X	X
RT7947	TH-2, 100	Thermistor, SDT-50	X	X	X
RT2334	D-1	ALC Diode, 1S188	X	X	X
RT7946	D2, 100	Governor Diode, 1S1212	X	X	X
		COILS & TRANSFORMERS			
RT6580	L-1	Bias Trap Coil	X	X	X
RT6581	L-2, 3,100, 101	Motor Choke Coil	X	X	X
RT6579	T-1	Bias Oscillator Transformer	X		
RT8513	T-1	Bias Oscillator Transformer		X	
*EA35X32	T-1	Bias Oscillator Transformer			X

CAT. NO.	REF. NO.	DESCRIPTION	VERSION A	B	C
		MISCELLANEOUS			
RT6578	SW-1	Record/Play Switch	X	X	X
RT6585		Built-in Mike Element	X		
RT8515		Built-in Mike Element		X	
*EA62X40		Built-in Mike Element			X
RT6577		Power Switch	X	X	X
RT6576		Jack Assy	X	X	X
RT7964		Positive Battery Contact	X	X	X
RT7965		Negative Battery Contact	X	X	X
RT9858		Amplifier Board w/Comp	X		
RT9995		Amplifier Board w/Comp		X	
*EA93X131		Amplifier Board w/Comp			X
RT9996		Governor Board w/Comp	X		
RT9859		Governor Board w/Comp		X	
*EA93X132		Governor Board w/Comp			X
RT7966		Transistor Heatsink	X	X	
*EA2X97.		Transistor Heatsink			X
		ACCESSORY ITEMS			
RT6599		External Microphone	X	X	X
RT3657		Earphone	X	X	X
RB2909		Carrying Case	X	X	X
RT9543		AC Convertor	X	X	X
RT9058		Foot Pedal Switch	X	X	X
RT7882		Foot Pedal Switch Adaptor	X	X	X
RT9507		Patchcord Kit	X	X	X
RT7201		Blank C-60 Cassette	X	X	X
RT7220		Blank C-120 Cassette	X	X	X
RT7230		Head Cleaning Cassette	X	X	X
RT7817		Jewelers Screwdriver Set	X	X	X
		CABINET & APPEARANCE ITEMS			
RT6588		Battery Cover	X		
RT8516		Battery Cover		X	
RB2788		Cabinet Side	X		
RT8517		Cabinet Side		X	
*EA89X171		Cabinet Side			X
RT6593		FWD/RWD Button	X	X	X
RT6592		Record Button	X	X	X
RT6594		Play Button	X	X	X
RT6587		Cabinet Back Assy	X	X	X
RB 2789		Cabinet Top	X	X	X
RB2787		Cabinet Side	X	X	X
RB2790		Cabinet Bottom	X	X	X
RT9861		Chassis Panel	X	X	X
RT7957		Battery Case Assy	X	X	X
RT9862		Governor Board Cover	X	X	X
RT6589		Button Pad (Large)	X	X	X
RT6590		Button Pad (Small)	X	X	X
RT6586		Cassette Lid Assy	X	X	X
*EA2X94		Rubber Cover, Microphone	X		
*EA2X95		Rubber Cover, Microphone		X	
*EA2X96		Rubber Cover Microphone			X

Chapter 6

Cartridge Tape Transport Mechanisms: Service and Repair

Most cartridge tape (8 track) machines are in cars, but a few are used in home entertainment systems. The majority of your service and repair work will probably be done on equipment installed in automobiles.

Figure 6-1. A cartridge machine. (Courtesy of 3M)

Understanding mechanical principles of tape cartridge transport mechanisms

Cartridge players use only one motor—almost all machines run in one direction only—and there are no fast-forward or rewind modes. The tape runs through the machine continuously; a series of tracks is played as the tape moves by the heads. Four separate programs are recorded on each length of tape; the eight tracks are divided into the following four combinations of two tracks each: tracks 1 and 5, tracks 2 and 6, tracks 3 and 7, and tracks 4 and 8. The player then returns to tracks 1 and 5 after running through the series and starts through again.

Tape travels through the machine at 3-3/4 ips and is maintained at that speed by a capstan pressing it against the drive wheel. Tape head positioning can be done manually or automatically. Manual positioning is achieved by moving the selection lever. The general method used is that the switch is thus shorted, one end of the solenoid is grounded, the solenoid plunger returned to its original position, moving a cam to its next position which, in turn, moves the studs on the head, changing its position. Automatic positioning is done by a strip of conductive tape at the end of the program which shorts out the selection switch, thus starting the same action as would result if the switch were moved by hand.

The basic operational mode is *playback*, although some home entertainment systems record as well. Recording 8 track cartridges differs from recording open-reel or cassette tapes because the absence of *fast-forward, pause,* or *rewind* modes makes mechanical editing impossible and electronic editing difficult (see Chapter 10).

Servicing components in cartridge players is similar to servicing parts in open-reel or cassette players. Since fewer modes are used in 8 track players, finding the source of the trouble is often simpler.

The *cartridge* (a continuous loop of tape winding on a single reel) may jam or break. Customer complaints about malfunctioning equipment are often the result of defective cartridges. Worn or damaged pressure pads, sticking wheel hubs, warped or heat distorted cartridges will produce poor sound or will render the player inoperable.

The *capstan* (used to maintain steady tape speed past the

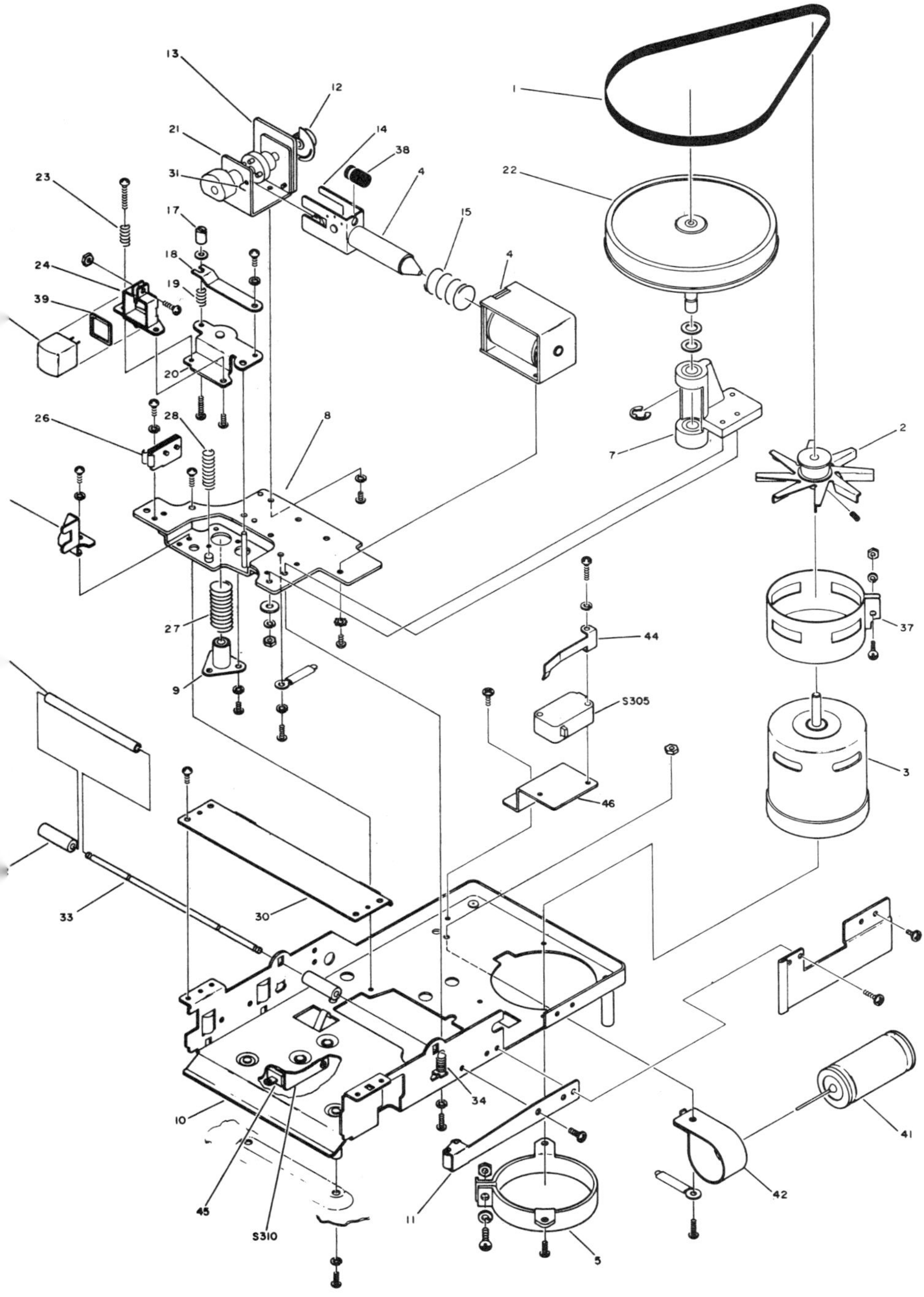

Figure 6-2. Exploded view of tape deck assembly. (Courtesy of General Electric)

heads) may cause trouble if it is bent or damaged. Although it may be possible to straighten the shaft or repair it temporarily, the best solution is to replace it.

Belt drives (usually flat, with the rough side on the inside) as well as other rubber parts deteriorate or break and must be replaced. You may clean them with denatured alcohol, but it is a good rule of thumb to replace any belt in a cartridge player that shows the slightest sign of wear; you may save yourself another dissassembly job before the player is returned to your customer. Your time is more valuable than a simple part.

Motors will be dc motors (in which the speed is changed by changes in voltage) in players used in cars, and hysteresis-synchronous motors (in which the speed is changed by current frequency) in high-quality home entertainment systems. Motors will seldom have to be replaced.

Most of your work with cartridge players will be in cleaning and adjusting parts to original specifications.

Although some newer equipment boasts 4 track or 4 channel playback (for quadraphonic sound), the basic mechanical principles are the same as in 8 track equipment. The major difference is the use of a 4 track head (which covers 4 tracks instead of 2) which must be moved two positions instead of four as with 8 track players.

Other models boast *fast-forward* modes (using a belt system to change pulley ratios) or a *record* mode (usually by adding a separate electronic system onto the conventional *playback* system). Such equipment is only slightly more difficult to service than more common 8 track models.

A handy checklist for spotting the cause of the trouble

Test the equipment with a cartridge you know to be good. Check to see the electrical connections are clean and tight. After making sure that the malfunction is not due to external reasons, do one (or more) of the following to correct the trouble.

1. *If the motor doesn't run*
   a. replace the blown fuse (car models).
   b. correct the poor contact of the power supply cord.
   c. repair or replace the defective micro-switch.
   d. replace the open filter choke or choke coil.
   e. correct the faulty ground connection (car models).

f. replace the defective capacitor at the motor.
g. replace the motor.

2. *If the capstan doesn't turn*
   a. replace the broken drive belt.
   b. tighten the loose screw on the pulley (if needed).
   c. replace the motor.

3. *If the tape starts too slowly*
   a. use a different (and dependable) cartridge.
   b. replace the defective drive belt.
   c. clean the greasy pinch roller, pulley, belt, flywheel or capstan.
   d. replace or tighten the capstan housing.
   e. replace the bent or damaged drive shaft.
   f. replace the defective flywheel.
   g. replace the defective oilless bearing (if any).
   h. replace the defective motor.

4. *If the speed is too fast*
   a. replace the defective motor governor.

5. *If the tape speed is too slow*
   a. use a different (and dependable) cartridge.
   b. stop the belt from slipping.
   c. stop the pressure roller from slipping.
   d. replace the defective drive shaft.
   e. replace the defective flywheel.
   f. repair or replace the defective capstan.
   g. replace the defective capstan housing thrust cap.
   h. replace the motor.

6. *If the tape breaks or jams*
   a. use a dependable cartridge.

7. *If the tape is not taken up*
   a. replace the defective cartridge.
   b. adjust the reel base tension to specifications (usually 55 to 70 grams per centimeter).
   c. stop the drive belt from slipping.
   d. correct the malfunctioning idler.

8. *If fast-forward does not operate*
   a. replace the defective cartridge.
   b. stop the drive belt from slipping.
   c. replace the broken belt.
   d. repair or replace the inoperative leaf switch.

9. *If wow-and-flutter occurs*
   a. replace the defective cartridge.
   b. replace defective belts or drive wheels

c. align the heads.
d. lubricate the flywheel bearing.
e. adjust the pinch roller spring to specifications (usually about 40 grams).
f. replace the defective flywheel and housing.
g. replace faulty take-up mechanism.
h. balance the motor.
i. replace the defective motor.
(Wow-and-flutter is usually difficult to trace and correct; thus, you may need to perform several of these operations.)

10. *If the program doesn't change*
a. replace the defective cartridge (which probably has a defective conductive tape).
b. correct the pad pressure on the head (which is too weak).
c. replace the defective plunger.
d. replace the defective push switch.
e. replace the defective resistor in the circuit.

11. *If the program pilot light doesn't work*
a. clean or replace switch.
b. clean or replace selector assembly and contacts.

12. *If the program switches in error during playback*
a. adjust terminal contacts.

13. *If head indexes incorrectly*
a. increase tension on solenoid plunger return spring (or replace it).
b. replace or increase the tension of the cam tension spring.
c. replace the defective cam and shaft assembly.
d. adjust the solenoid plunger-end stop screw. (Turn the screw clockwise until the tape head stops indexing, then turn the screw counter-clockwise one-half to one turn. Fasten the lock nut.)

At times you will find it necessary to adjust the tape head up or down (*height adjustment*) or to move it at an angle (*azimuth adjustment*).

Use a commercial test tape (instructions will vary). In general, the following procedure and specifications will be adequate when accurate specifications are unavailable.

For correct *height adjustment* (misadjustment causes cross-talk or poor frequency response), set the output level control or *balance* at maximum for the left channel. Attach an oscilloscope or VTVM across the left channel. Loosen the hex nut (if present) and adjust the screw for a NULL reading (see illustration). Always make the height adjustment first.

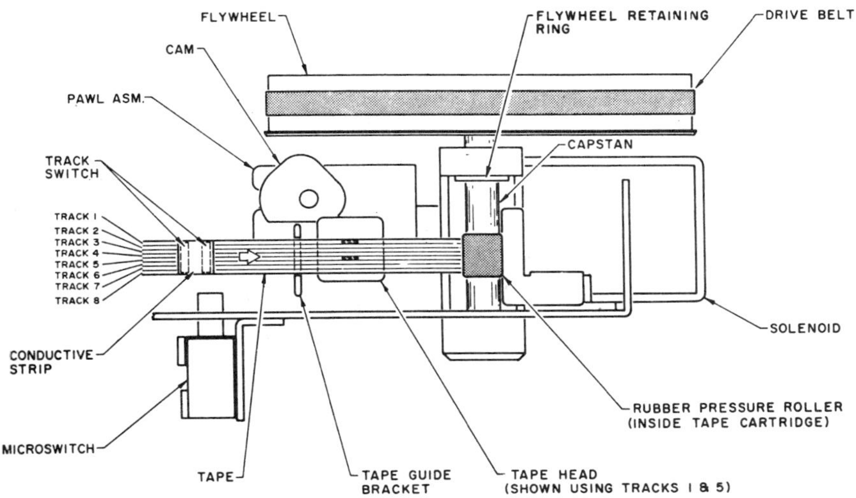

Figure 6-3. How the head moves across an 8-track tape. (Courtesy of General Electric)

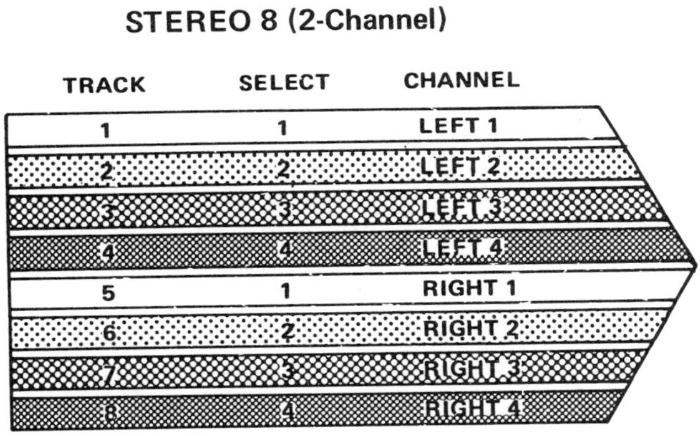

Figure 6-4. Track location (Courtesy of 3M)

For *azimuth adjustment* (misadjustment causes poor high frequency response), attach the oscilloscope or a VTVM to the right channel. Turn the output level or *balance* control to the maximum for the right channel. Loosen the hex nut (if present) and adjust the screw for MAXIMUM output.

Generally, the program selector will be on *program 2* (tracks 2 and 6). Do not change the program selection but repeat the adjustment several times (one adjustment may throw the other off). Seal the screws with glyptal (a nonhardening cement).

Some models have a third adjustment—*tilt*. The purpose of this adjustment is to maintain uniform contact with the tape against

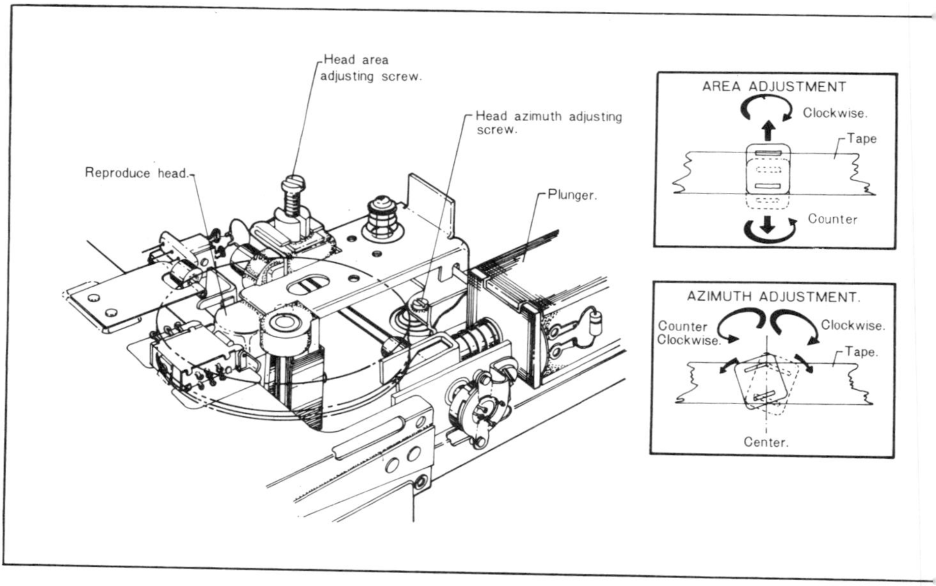

Figure 6-5. How to adjust height and azimuth (Courtesy of Clarion Corporation of America, Los Angeles, California)

the head to assure equal output from all channels—especially important with four-channel tapes for quadraphonic sound. You need a special test tape (which contains an 8000 Hz signal recorded on the full width of the tape). Adjust the *tilt* screw for equal output from each channel.

As part of your standard servicing procedures, clean all drive surfaces. You may need to adjust *torque* according to specifications—usually about 50 grams for the pinch roller. Finally, demagnetize the heads as described subsequently.

Useful short-cuts for in-car servicing

Before removing the player from the car, stop and think. You may save yourself hours of fruitless labor—the problem may not be in the equipment itself. Instead, the customer's complaint may be the result of a damaged speaker (particularly if the speakers are mounted in the doors, where they are subject to much abuse), a loose ground (check to see it is clean and tight) or noise from the alternator or spark plugs (suggest a tuneup or install suppression devices).

Very often, the equipment needs no more than a good cleaning (oxide buildup on the heads ruins good sound) or demagnetiza-

Figure 6-6. One of many useful servicing cartridges. (Courtesy of Ampex)

tion (magnetized heads cause noisy playback). You can save time and service the equipment in the car by using demagnetizers and tape cleaners which come in cartridge form. These effort savers need only to be inserted in the player. There are various combinations available: alignment/cleaner cartridges, demagnetizer/cleaner cartridges. You can thus do two jobs at once and build your profits.

How to improve playing quality by periodic maintenance

You can often dramatically improve the sound quality of a cartridge player by simple maintenance procedures. Sell your customers on a tuneup every six months. You can clean and demagnetize the equipment without taking it out of the car. For a better job, pull the equipment and clean it on your bench with a soft cotton swab and alcohol (which will quickly remove the oxide buildup from the heads). Since magnetized heads cause a loss of high frequency response and increase the noise level, use a demagnetizer (pass it over the heads and slowly withdraw it according to instructions). Adjust the head height and azimuth if necessary. At the same time you might notice some problem that your customer has overlooked—thus bringing you additional profits for its repair, and making your customer pleased that he has avoided an extra trip to the repair shop.

REALISTIC®

Service Manual

14-1920

STEREO 8 PLAYER

TR-167

Catalog Number : 14-1920

CUSTOM MANUFACTURED FOR RADIO SHACK TC A TANDY CORPORATION COMPANY

SPECIFICATIONS

GENERAL SPECIFICATIONS		
DESCRIPTION	AC operated 8 track cartridge tape player with pre-amplifiers.	
TAPE	Tape drive	Capstan drive
	Tape speed	3-¾ inches per second
	Play	8 track dual channel play
CONTROL & FUNCTIONS	Program selector	Push button and automatic
	Line output level control	Rotary—on back
INDICATOR	Program indicator	Numbered program to be lighted
EXTERNAL CONNECTIONS	AC wire	8 foot power supply cord with plug
	Line output	2-4 foot output leads with RCA type plugs
POWER SUPPLY	AC 120 V 60 Hz	
MOTOR	Rated voltage	DC 21V
	Rated speed	3000 RPM.
	Rated current	220 mA
	Rated torque	25 g—cm
POLARITY	Negative ground	
OUTPUT LEVEL	Line out	0—600 mV (100 kohm load)
CIRCUIT	6 Transistor 3 diode	

TAPE PERFORMANCE SPECIFICATIONS				
TEST ITEM	UNIT	NOMINAL	LIMIT	TEST TAPE
1.1 SRL output level	mV	900	600	VTT807
2.1 Distortion at playback	%	1.5	3	VTT807
3.1 S/N ratio	dB	45	40	VTT807
6.1 Separation	dB	43	38	VTT819
6.2 Track crosstalk	dB	45	40	VTT804/820
7.1 Audio fidelity at playback 50 Hz(1kHz=0db)	dB	−2	+4, −6	VTT808/816
8kHz	dB	0	+6, −6	VTT808/816
Tape speed deviation	%	0	+3, −2	VTT802
Wow & Flutter (JIS RMS)	%	0.25	0.35	VTT802

TEST CONDITION	TAPE: Victor Test Tape or equivalent to use Output terminal resistive load: 100 kohm Test voltage: AC 120V 60 Hz Test output level: 500 mV
OUTPUT LEADS	Red — Left Gray— Right

NOTE: Nominal Specs represent the design specs; all units should be able to approximate these--some will exceed and some may drop slightly below these specs. Limit Specs represent the absolute worst condition which still might be considered acceptable; in no case should a unit perform to less than within any Limit Spec.

DISASSEMBLY INSTRUCTIONS

1. Remove 4 screws (E) from the rear of the cabinet.
2. Remove 4 screws (A), 4 washers (B), 4 feet (C), and 4 chassis mounting washers (D) from the bottom of the cabinet.

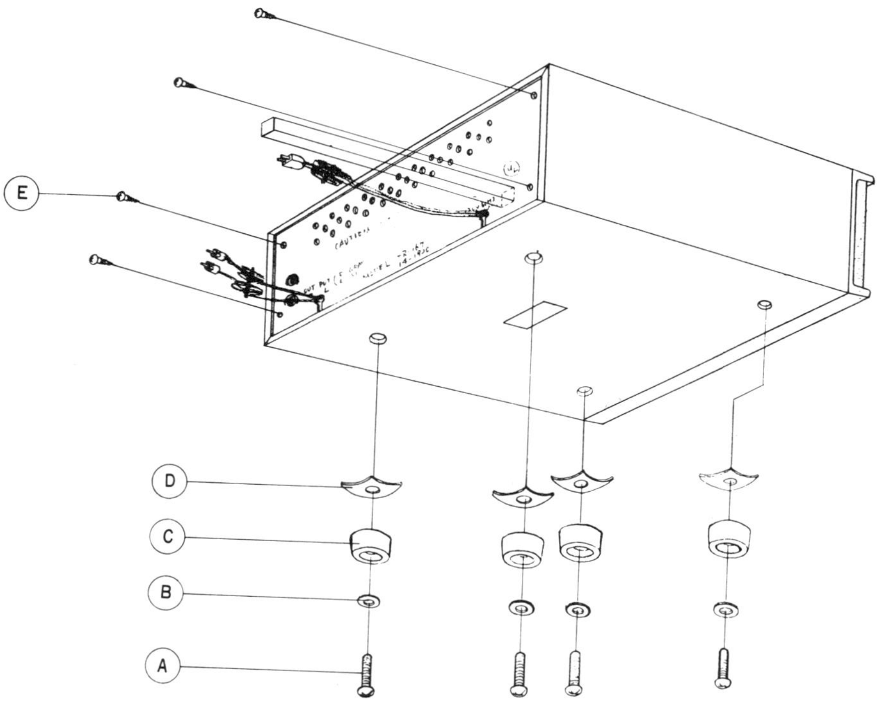

HEAD CLEANING INSTRUCTIONS

1. Clean head with cotton swab and head cleaner or denatured alcohol.
2. Wipe dry with soft dry cloth.

HEAD AZIMUTH AND HEIGHT ADJUSTMENT

Before starting this adjustment, demagnetize the head and all tools to be used.
Set OUTPUT LEVEL controls to "max"
The following equipment is required for head adjustment:

1. VTVM
2. Load resistor (100 kohm)
3. AC power source (AC 120 V 60 Hz)
4. Test tape cartridge RCA #321
5. Small screwdriver
6. Scope (optional)

1. Connect the VTVM and load resistor to right and left output as shown in Figure 1.
2. Select program 2 by depressing program select button (head making contact with tracks 2 and 6).
3. Turn the height adjust screw (Figure 2) back and forth to locate the position at which left output is minimum.
4. After height adjustment has been completed, turn the azimuth adjust screw (Figure 2) to locate the position at which right channel output is maximum.
5. Repeat these adjustments (Items 3 and 4) alternately, since height may change by turning the azimuth screw.
6. After all adjustments have been completed, secure the height adjusting and azimuth screws by applying glyptal or other non-hardening cement to their heads.

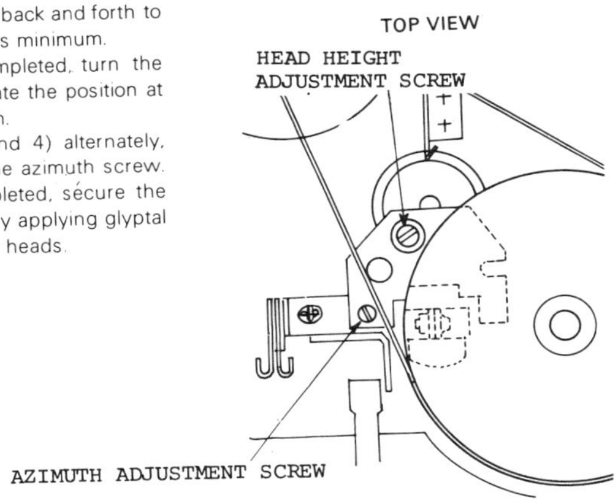

Figure 2. Adjustment Locations

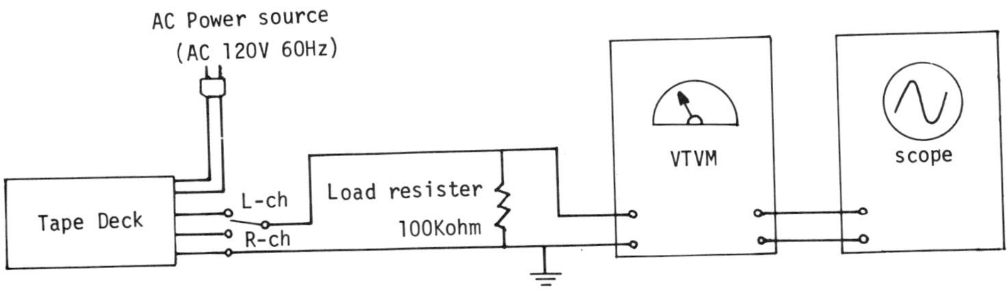

Figure 1. Adjustment Setup

TROUBLE SHOOTING GUIDE

CIRCUIT	TROUBLE	CAUSE	REMEDY
Pre-Amplifier	No sound	Defective 2SC828, 2SC644 (TR1-6). Defective C3, 4, 13, 14, 17, or 18. Open or Short Circuit (lead), Head (L. Head, R. Head)	Replace Replace Replace
	Poor Sound Volume	Oxide Deposite on Head (L. Head, R. Head). Defective 2SC828, 2SC644 (TR1-6)	Clean Replace
	Hight Pitched Tone	Incorrect Head Azimuth Angle. Defective Head. Defective Capacitor (C1, C2) or Open Capacitor Lead	Adjust Replace Replace
	Low Output	Defective Capacitor C7, 8. Defective R13, 14. Defective R5, 6.	Replace Replace Replace
	Distorted Output	Defective R1, 2, 7, 8, 11, 12, 15, 16, 17, 18, 19 or 20.	Replace
	Abnormal Oscillation	Head Lead Defective.	Replace
Solenoid Circuit	Head Switching Impossible	Defective Solenoid Switch Contact. Shorted Diode (D-1). Solenoid Open or Layer Short.	Replace Replace Replace
Power Source	No Sound	AC-Cord Open or Break. Shorted Diode (D-2, 3)	Replace Replace

TRANSISTOR LEAD IDENTIFICATION

TR1, TR2 2SC644

TR3, TR4, TR5, TR6 2SC828

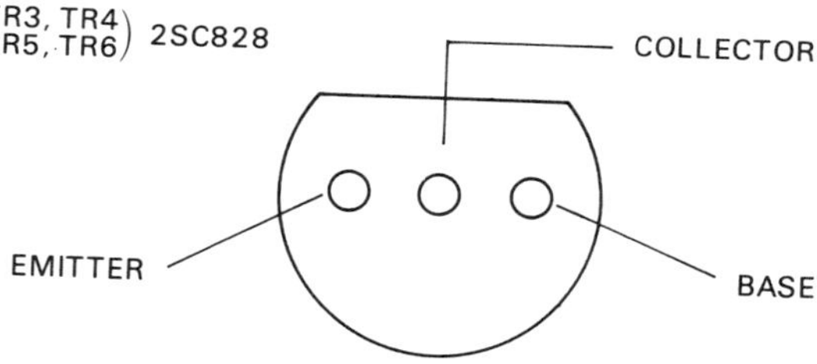

BOTTOM VIEW OF TRANSISTOR

POWER SUPPLY P.C.B. (TOP VIEW)

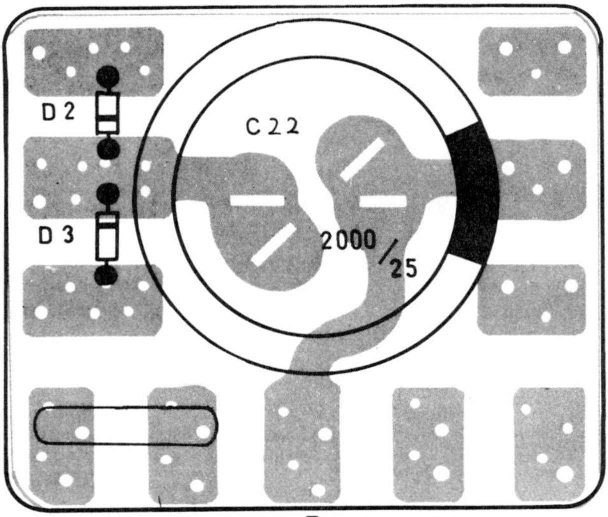

C 24 .0047uF 150VAC

POWER SUPPLY P.C.B. (BOTTOM VIEW)

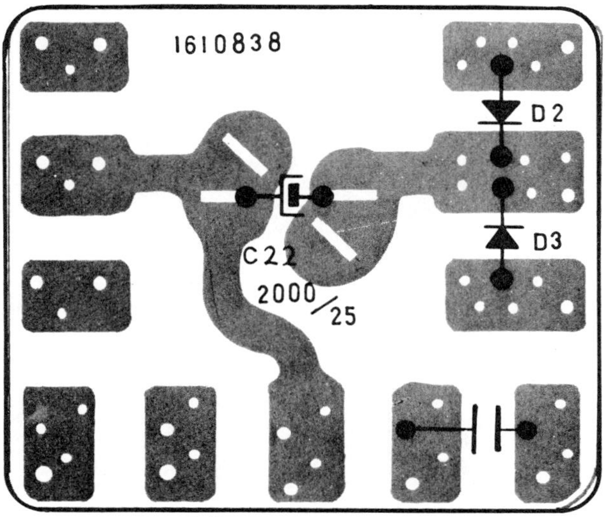

C 24 .0047uF 150VAC

PRE-AMPLIFIER P.C.B. (TOP VIEW)

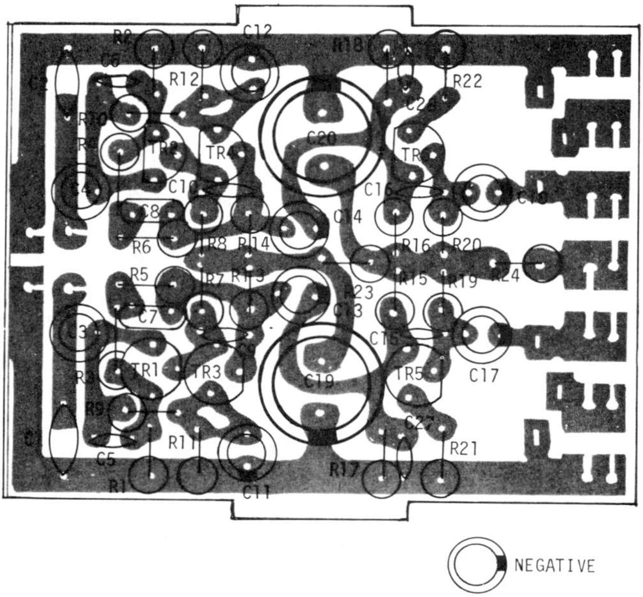

PRE-AMPLIFIER P.C.B. (BOTTOM VIEW)

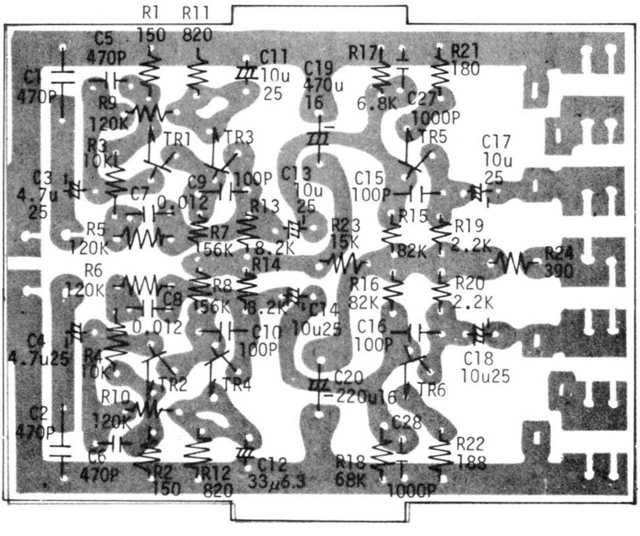

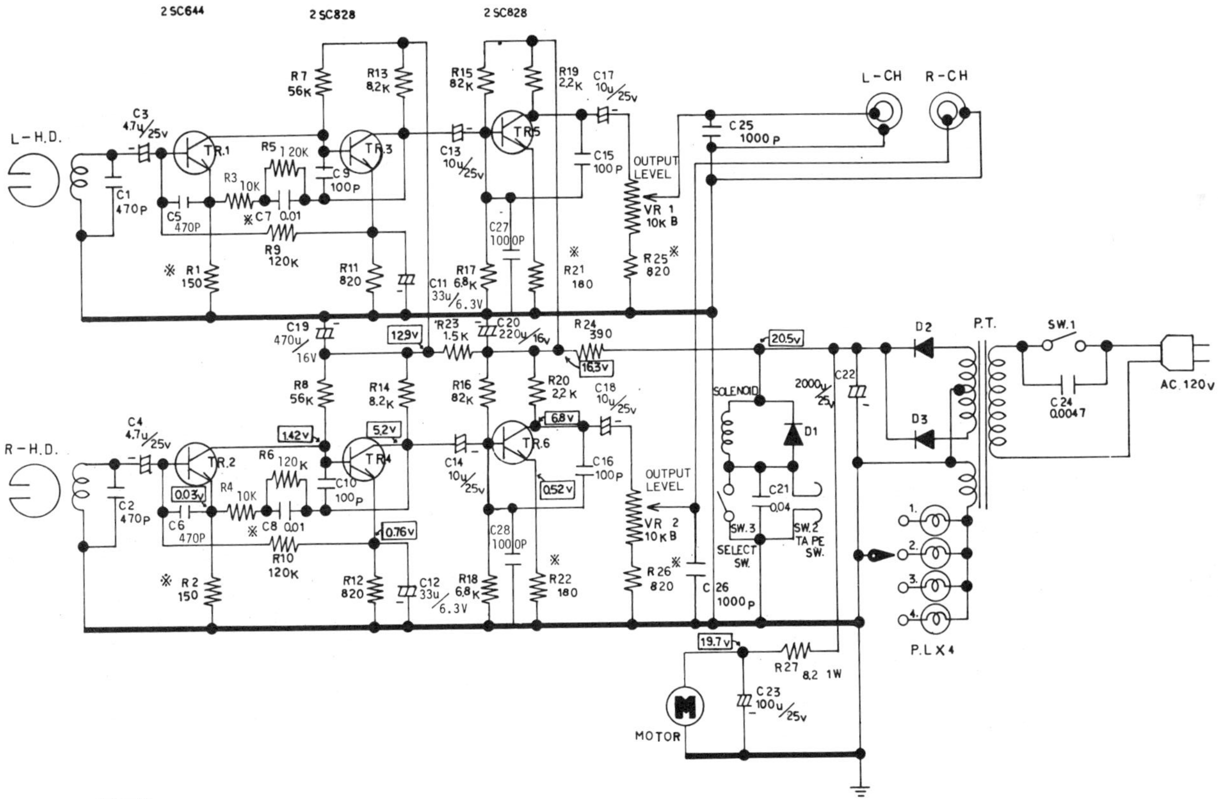

NOTES

※ Typical value and varies from unit to unit.
Resistance values are indicated in ohms (K = 1,000 ohms).
Capacitance values are indicated in μF (P = $\mu\mu$F).

WIRING DIAGRAM

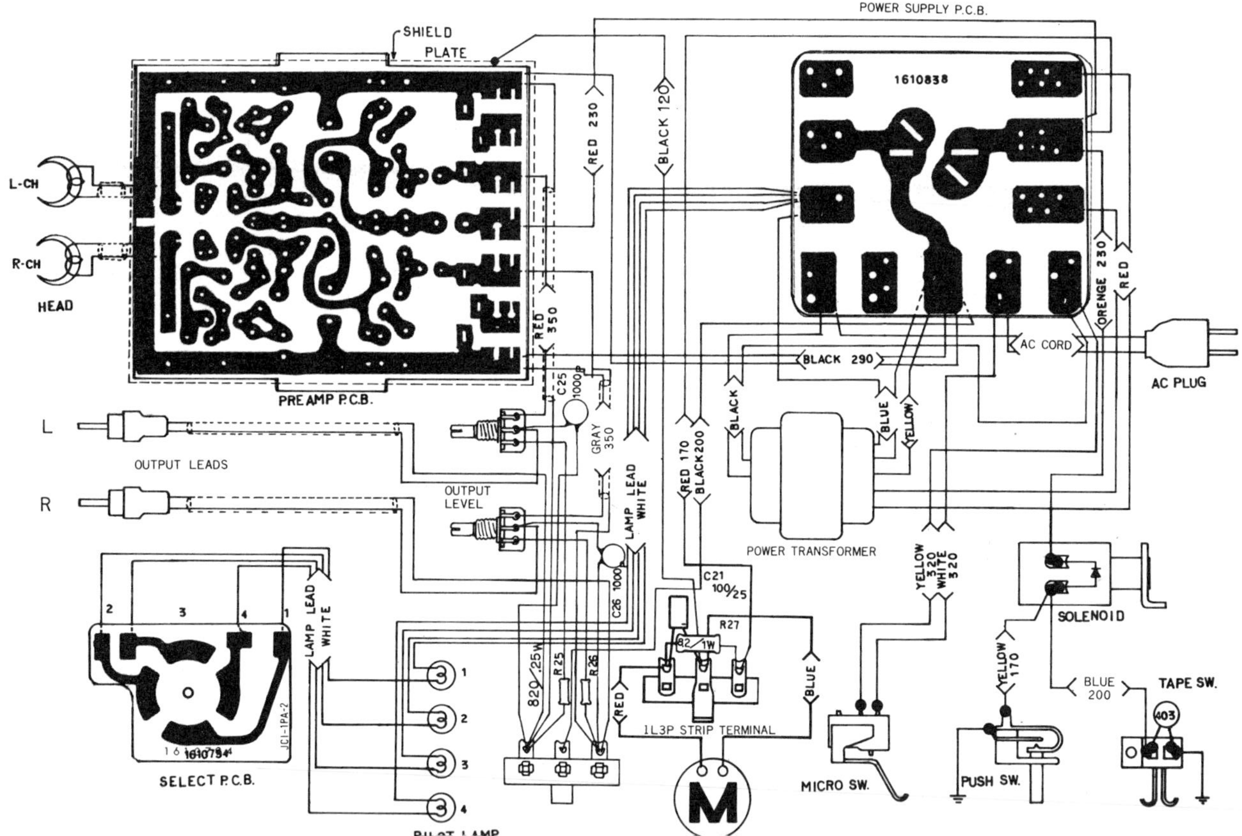

EXPLODED VIEW

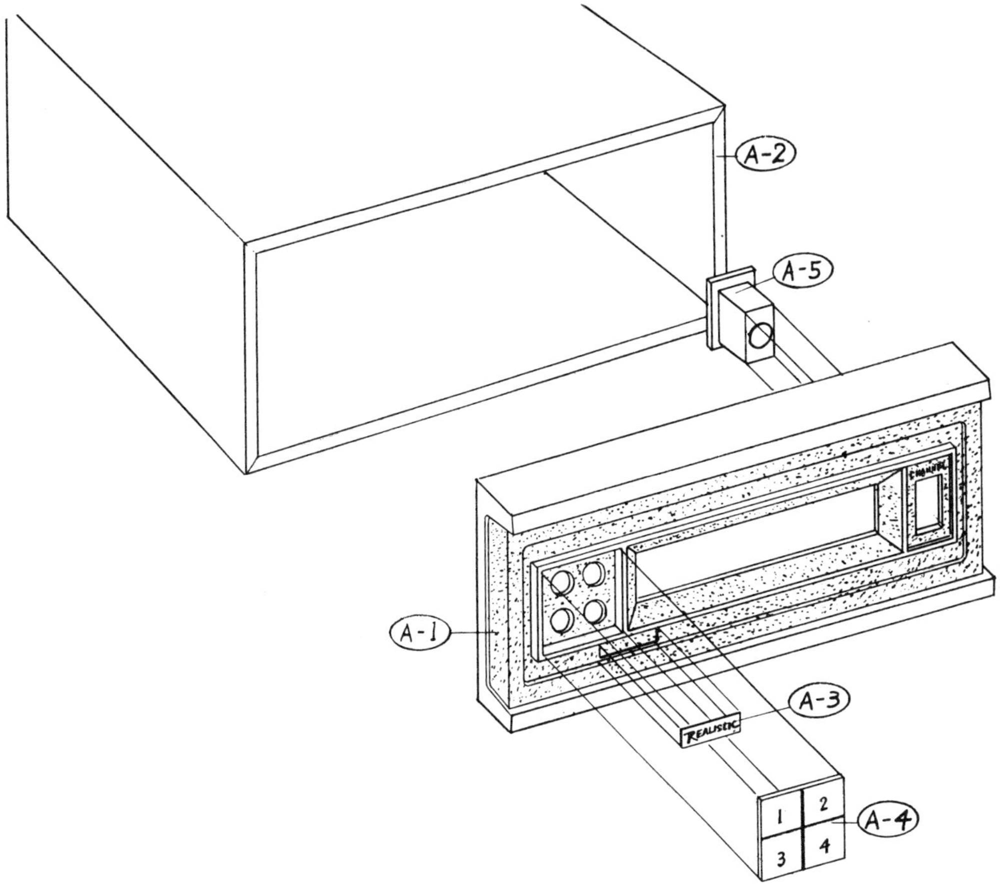

EXPLODED VIEW

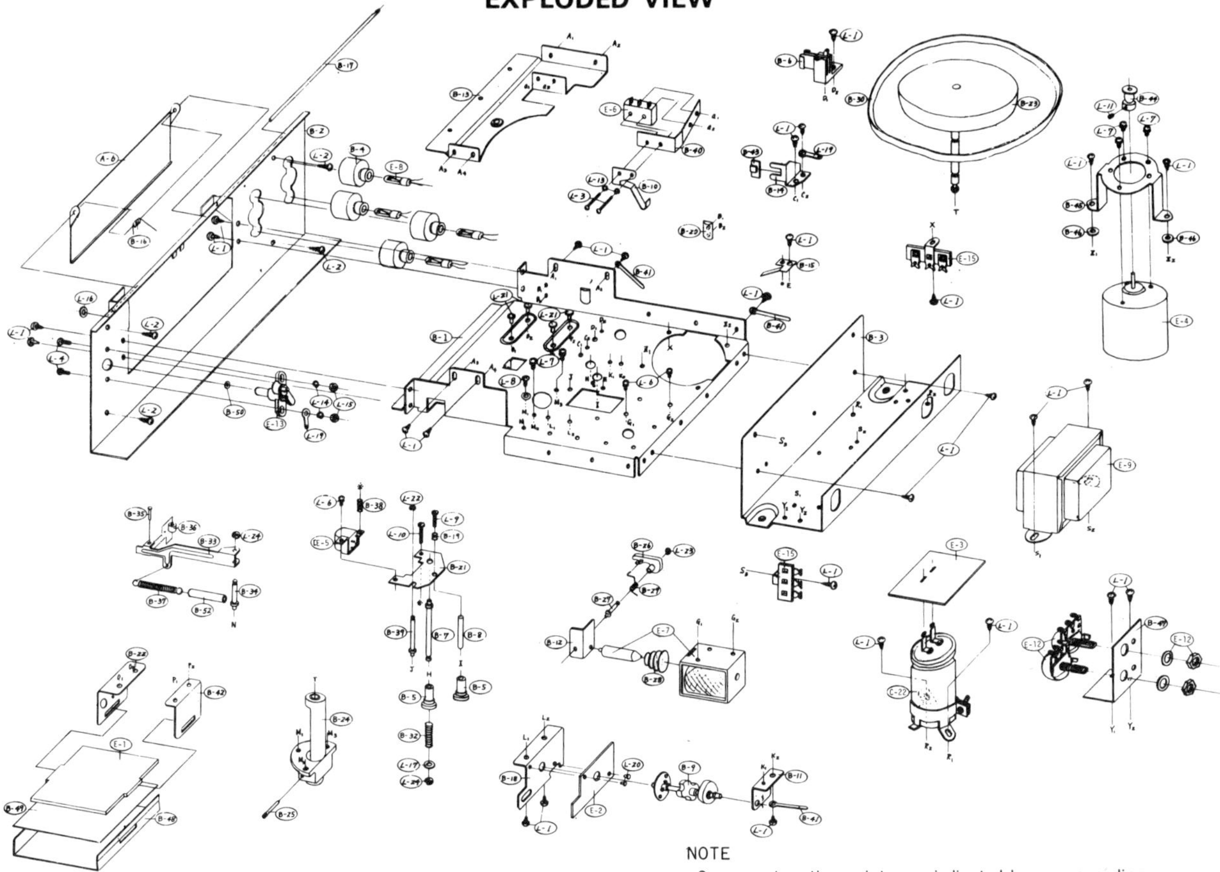

NOTE

Component mating points are indicated by corresponding letters. For example: power transformer (E-9) mounting holes S_1, S_2 mate with deck stand (B-3) mounting holes

REPLACEMENT PARTS LIST

ELECTRICAL PARTS

Symbol No	RS No	Description	Mfr No
C1,2, 5,6		*Ceramic capacitor 470 pF	1271471
C3,4		Electrolytic capacitor 4.7μF,25V	1204475
C7,8		*Mylar capacitor 0.01μF	1252103
C9,10, 15,16		Ceramic capacitor 100pF	1271101
C11,12		Electrolytic capacitor 33μF,6.3V	1201336
C13,14, 17,18		Electrolytic capacitor 10μF,25V	1204106
C19		Electrolytic capacitor 470μF,16V	1203477
C20		Electrolytic capacitor 220μF,16V	1203227
C21		Ceramic capacitor 0.04μF	1273403
C22		Electrolytic capacitor 2000μF,25V	1220045
C23		Electrolytic capacitor 100μF,25V	1204107
C24		Ceramic capacitor 0.0047μF,150V AC	1220097
C25,26, 27,28		Ceramic capacitor 1000pF	1271102
E-1	X-4754	Pre-amp P.C.B.	1610461
E-2	X-4755	Selector SW P.C.B.	1610794
E-3	X-4756	Rectifier P.C.B.	1610838
E-4	M-4277	Motor	164N046
E-5	H-4230	Playback head	1650027 1650040 1650050
E-6	S-6031	Micro switch (SW1)	1623074 1623021
E-7		Solenoid	166Y030
E-8	L-0504	Pilot lamp	1830096

Symbol No	RS No	Description	Mfr No
E-9	TA-0436	Power transformer (P.T.)	118G183A 118G183B
E-10	W-1721	Power supply cord	1750172
E-11	W-1722	Patch cord	1750176
E-12	P-0748	Variable resistor 10kohm B (VR½)	139N097
E-13		Push switch (SW3)	1623060
E-14	HB-1000	Bushing	CHSR3P4
E-15	HB-1001	Lug (1L2P)	1720064
D1,2, 3		Diode	SR1K-2 V06, 10D-1
R1,2		*Carbon resistor 150ohm	1337151
R3,4		Carbon resistor 5.6kohm	1337562
R7,8		Carbon resistor 56kohm	
R9,10		Carbon resistor 120kohm	1337563
R11,12		Carbon resistor 820ohm	1337124
R13,14		Carbon resistor 8.2kohm	1337821
R15,16		Carbon resistor 82kohm	
R17,18		Carbon resistor 6.8kohm	1337823
R19,20		Carbon resistor 2.2kohm	1337682
R21,22		*Carbon resistor 180ohm	1337222
R23		Carbon resistor 1.5kohm	1337181
R24		Carbon resistor 390ohm	1337391
R25,26		*Carbon resistor 820ohm	1326821
R27		Metal oxide film resistor 8.2ohm, 1W	1344829
Tr1,2		Transistor	2SC644FR
Tr3,4		Transistor	2SC828FR
Tr5,6		Transistor	2SC828FR

MECHANICAL PARTS

Symbol No	RS No	Description	Mfr No	Symbol No	RS No	Description	Mfr No
A-1	Z-1953	Front	21F7062	B-23	RA-3191	Flywheel	25R7034
A-2	Z-1954	Wood Cabinet	27C7079	B-24	RA-3192	Flywheel Metal	22R7012
A-3	HB-0979	Brand Badge	23M7458	B-25	HB-0987	Stop Pin	25W0444
A-4	HB-0980	Indicator Window	24D7044	B-26	RB-4873	Channel Selector Arm	21W0191
A-5	K-1625	Channel Button	21N7212	B-27	RB-4874	3 ϕ Shaft	25W1306
A-6	HB-0981	Cartridge Door	23M7459	B-28	RB-5099	Solenoid Spring	26W7079
A-7	Z-1955	Back Lid	24C7059	B-29	RB-5100	Selector Arm Spring	26W0185
A-8	F-0147	Foot	21C7049	B-30	B-6254	Neoprene Belt	21V7020
A-9		Caution Label	24L4022	B-32	RB-5101	Cam Pressure Spring	26W7136
A-10	HW-0812	Chassis Mtg. Washer	23W0478	B-33	RB-4875	Pressure Arm	23W1145
A-11		Serial No. Label	24L8511	B-34	RB-4876	Pressure Arm Shaft	25W1035
B-1	Z-1956	Main Chassis	23S7099	B-35	RB-4877	Pressure Roller Shaft	25W0741
B-2	Z-1957	Deck Front	23S7174	B-36	RA-0137	Pressure Roller	21W0054
B-3	Z-1958	Deck Stand	23S7133	B-37	RB-5102	Pressure Spring	26W0111
B-4	HB-0982	Lamp Holder	21W7127	B-38	RB-5103	Slit Adj. Spring	26W7006
B-5	RB-4865	Shaft Boss	25W7458	B-39	RB-4878	Guide Shaft	25W7194
B-6	S-0613	Tape Switch Assy. (SW2)	21N7170	B-40	HB-0988	Micro Switch Fiber	24T7053
B-7	RB-4866	Head Setting Plate Shaft	25W7459	B-41	HB-0989	Wire Clamper	23W1046
B-8	RB-4867	Basic Shaft	25W7460	B-42	HB-0990	P.C.B. Holder B	23W7045
B-9	RB-4868	Cam Shaft Assy	22R7040	B-43	RB-4879	Tape Leader	21W7126
B-10	RB-5096	Micro Switch Spring	23W7833	B-44	RB-2431	Motor Pulley	25R7035
B-11	HB-1002	Cam Shaft Holder	23W7836	B-44	HB-0991	Motor Holder	23W8040
B-12	RB-4882	Solenoid Arm	23W7838	B-46	HW-0813	Rubber Washer	24W7271
B-13	HB0983	Cartridge Support	23W7832	B-47	RB-4880	Volume Setting Arm	23W8092
B-14	RB-4869	Tape Guide	23W7835	B-48	RB-4881	Shield Plate	23W8444
B-15	HB-0984	Cam Stopper	23W7837	B-49	HB-0992	Shield Fiber	24T7138
B-16	RB-5097	Door Spring	26W7134	B-50	HW-0814	Fiber Washer	24W8552
B-17	RB-4870	Door Shaft	26W7132	B-51		Schematic Diagram	24L7229
B-18	HB-0985	Selector P.C.B. Mtg. Plate	23W8081	B-52		Tube 6ϕ×30 mm	
B-19	RB-5098	Height Adj. Spring	26W7171	B-53	HB-0993	Indicator Acryl	24D7057
B-20	RB-4871	Cartridge Guide	21N7169				
B-21	RB-4872	Head Setting Plate	23W7834				
B-22	HB-0986	P.C.B. Holder A	23W7843				

SCREW & ETC.

Symbol No	RS No	Description		Mfr No	Symbol No	RS No	Description		Mfr No
L-1	HS-1561	Tapping Screw	M3×6	SZM3306	L-13	HW-0815	Spring Washer	M2.3	WSMY055
L-2	HS-1562	Tapping Screw	M3×8	SZM3308	L-14	HW-0816	Spring Washer	M3	WSM3067
L-3	HS-1563	Pan Head Screw	M2.3×14	SPM3Y14	L-15	HN-0241	Hexagon Nut	M3	NHM3030
L-4	HS-1564	Pan Head Screw	M3×8	SPM3308	L-16	HN-0242	Push Nut	M2.4	NTKA024
L-5	HS-1565	Pan Head Screw	M4×14	SPM3414	L-17	HW-0817	Plain Washer	4.3ϕ×8ϕ×0.8t	WPB4088
L-6	HS-1566	Assembled Screw	M2.6×5	CPM1905	L-18	HW-0818	Plain Washer	4.3ϕ×10ϕ×0.8t	WPM4108
L-7	HS-1567	Assembled Screw	M3×6	CPM3306	L-19	HB-0994	Ground Lug	M3	HE03006
L-8	HS-1568	Flat Head Screw	M3×6	SDM3306	L-20	HB-0995	Rivet	2ϕ×4	RFD2004
L-9	HS-1569	Cylinder Head Screw	M2.6×10	SAN1910	L-21	HB-0996	Rivet	3ϕ×6	RFA3006
L-10	HS-1570	Cylinder Head Screw	M2.6×12	SAN1912	L-22	HB-0997	E Ring	2ϕ	EEU0020
L-11	HS-1571	Slotted Set Screw	M2.6×4	SGM2904	L-23	HB-0998	E Ring	2.3ϕ	EEU0023
L-12	HS-1572	Wood Screw	M2.7×10	SWN1V10	L-24	HB-0999	E Ring	3ϕ	EEU0030

NOTE
* Typical value is shown and varies from unit to un

SERVICE MANUAL
MINI TYPE IC CAR STEREO
MODEL M-880

7715 DESMORE AVENUE, VAN NUYS, CALIFORNIA 91406.

MUNTZ STEREO

MUNTZ STEREO CORPORATION OF AMERICA. PHONE NO. (213) 989-5000.

∗SPECIFICATIONS

Reproduction:	8 track, 4 program, 2 channel
Tape speed:	3.75 ips. (9.5 cm/sec)
Wow and flutter:	Better than 0.3%
S/N ratio:	More than 45dB
Power amplitude:	More than 100dB
Crosstalk:	More than 30dB (for adjacent channel) More than 40dB (for adjacent track)
Reproduction frequency range:	50～10,000Hz
Load impedance:	4 Ω x 2
Power output:	More than 3.5W (for 10% distortion) More than 6 W x 2 (for Max. Volume)
Power supply voltage:	DC.14V (10.8V～15.6V), negative ground
Power consumption:	Less than 1.2A (at Max. output) Less than 3.5A (at plunger operated)
Semiconductors:	8 transistors, 1 diode and 2 IC Pre amplifier ..... TA-7063P (051-0011-00) x 2 (IC) AF amplifier ...... 2SC373 x 2 Driver amplifier... 2SC735 x 2 Power amplifier .. 2SB461 x 4 Transient suppressor ...... 1S1943 x 1
Weight:	1.6Kg (3.5lbs)
Dimensions:	Width: 140 m/m (5.5") Height: 55 m/m (2.16") Depth: 177mm (7.0") Effective depth: 5.9" (150 mm)

∗COMPONENTS

PU-4272-02	Car stereo unit	1 Set
280-3008-00	Owner's guide	1 Each
281-0402-00	Warranty card	1 Each
300-0490-00	Rear mounting bracket	1 Each
300-5074-00	Mounting bracket	1 Each
852-4649-00	Extension lead	1 Each
921-5701-00	Part bag	1 Set
950-4616-00	Packaging kit	1 Set

* FEATURES:

o **Mini car stereo**
Very compact, light weight, 8 track and 4 program reproduction unit specially designed for automobiles.

o **Number lighting type indicator**
By the employment of number lighting type indicator, the reproduction program in use can be identified at a glance.

o **ICs in pre-amplifier section**
IC's provide reproduction of powerful and beautiful sound and improvement of stability and reliability.

o New mechanism
Since the vertical method is employed in the head shift mechanism, adjustment is simple and expansion of high bands and stability of left and right outputs are improved.
A very simplified mechanism is used. Therefore, the unit is highly compact. Durability of the unit and stability of wow and flutter are improved by supporting a flywheel with a capstan housing special washer.

* INSTALLATION OF MAIN UNIT: (Refer to Fig. 1)

Install mounting bracket under instrement panel or under glove box by drilling 1/ 5 inch diameter holes. Use hex bolt (W-sems type) flat washer which are contained in the accessory bag.
Fix unit with perforated rear bracket, if necessary.

Note 1: If necessary, the main unit mounting bracket can be bent or cut according to the shape of installation location. Since mounting holes are provided in the side of the bracket at 10 mm intervals, the main unit can be installed on different levels.

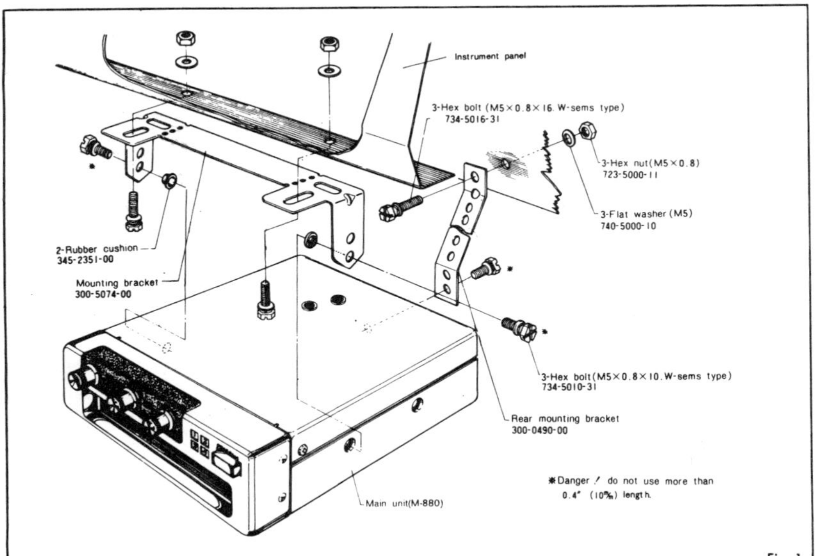

Fig. 1.

Note 2: You may install this unit within 45° angle.

* NEW MECHANISM:

1. **Complete block type and vertical-type head shift mechanism**

The vertical shift mechanism employs a method of shifting the head with the head face maintained vertically, which is different it from the expansion method of shifting the head by rotary motion around the center of the head arm fulcrum. Therefore, the allows for expension of high band sound and stability of left and right output.

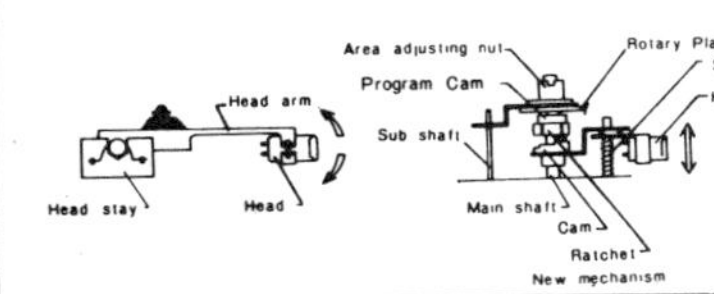

Fig. 2.

2. **Installation of flywheel** (Refer to Fig. 3)

Since the capstan housing supports the flywheel with special washers (poly-slider washers) between the flywheel and capstan housing, stability of wow and flutter and shock resistance are improved.

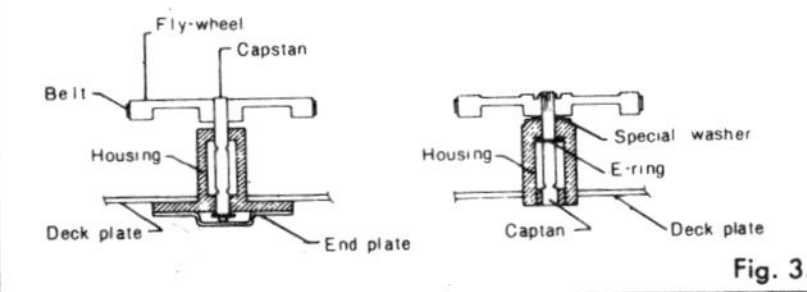

Fig. 3.

3. **Very small mechanism** (Refer to Fig. 4)

   Each individual "block" mechanism makes the entire mechanism small.

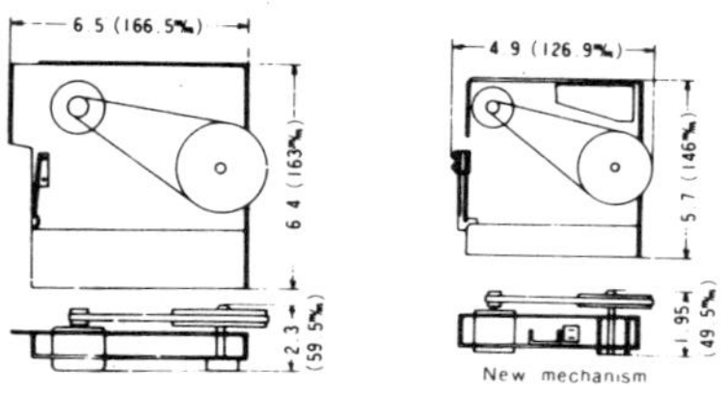

Fig. 4.

* ADVANTAGES AND FEATURES OF VERTICAL HEAD SHIFT MECHANISM:

1. **Completely unified head shift mechanism** (Refer to Fig. 5)

   All parts for the head shift mechanism; plunger, cam, ratchet, channel cam, channel lever, flywheel, etc. are provided in the die-cast housing section so that the head shift mechanism is very compact.

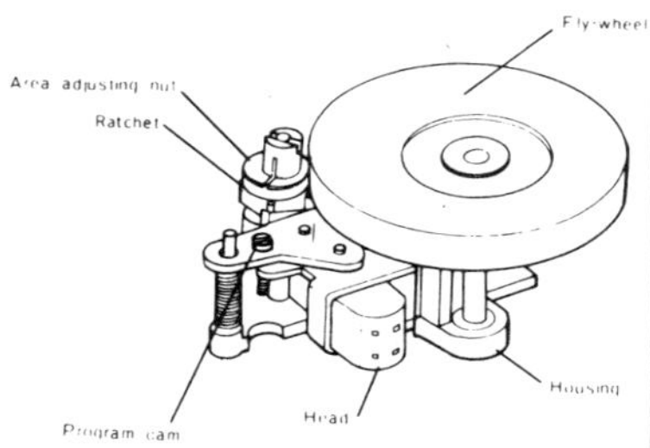

Fig. 5.

2. **Fixed cam and ratchet rotating method** (Refer to Fig. 6)

   This method which differs from the conventional method is Clarion's unique design, in which the cam is fixed and the ratchet is rotated.

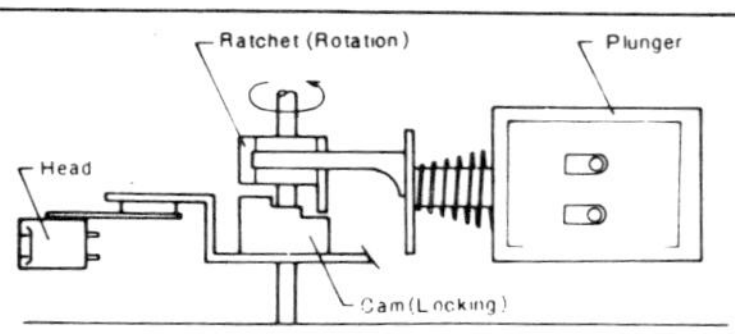

Fig. 6.

3. **Parts concentrated on shaft** (Refer to Fig. 7)

   Since an area adjusting screw as well as the small mechanism, cam, rachet, channel cam, rotary plate are all installed on one shaft, area adjustment is simpler and more accurate.

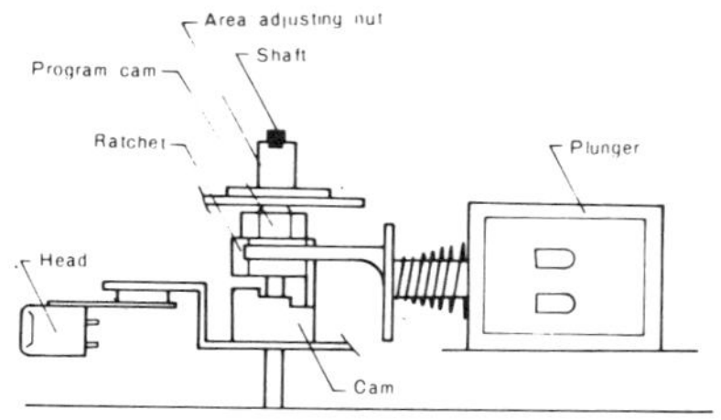

Fig. 7.

* ADJUSTMENT OF MECHANISM:

- **Adjustment of head location and angle**

  If the location and angle of the head is not correct in respect to the tape, there will be insufficient sound, poor tone or crosstalk. When these occur, perform adjustment by means of the area adjusting screw and azimuth adjusting screw on the rear of the head location.

- **Adjustment of head area** (Refer to Fig. 8)

  1. Balance left and right sides with the balance adjusting knob.
  2. Set the head location to the 2-6 track position (program indicator shows No. 2) with the test tape for area adjustment and adjust by turning the area adjusting screw to obtain maximum output.
  3. After the above steps, adjust the 1-5, 3-7, and 4-8 track positions in the same manner.

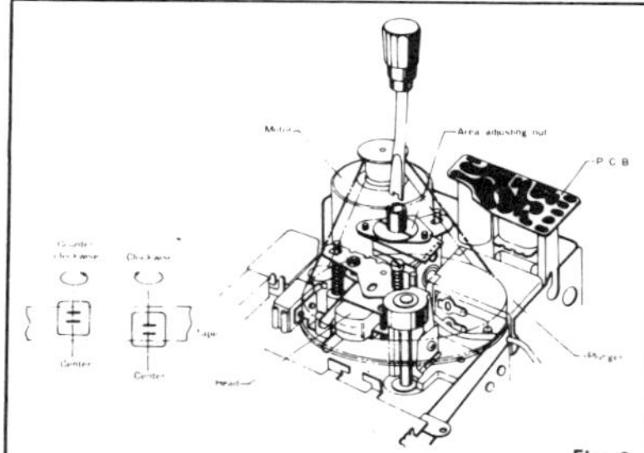

Fig. 8.

o **Adjustment of head azimuth** (Refer to Fig. 9)

1. **Set the location of the head to the 2-6 track position with the test tape for azimuth adjustment and adjust to obtain maximum output.**
2. **Area adjustment may deviate because of azimuth adjustment. Therefore, repeat azimuth and area adjustments several times.**

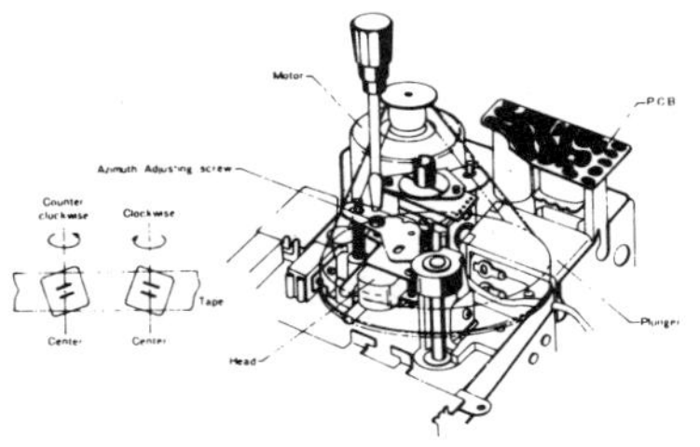

Fig. 9.

* IC

IC is the abbreviation for "Integrated circuits" containing transistors, resistors, capacitors on one base plate. They are used as one electric circuit. In other words, the IC is considered as one circuit unlike from individual parts such they transistors or resistors. Since ICs have various advantages, the have becomes rapidly popular in various devices.

* ADVANTAGES OF IC:

o **Improvement of operation efficiency**

By using ICs extensively, the number of parts is reduced, checking is simplified and troubleshooting and repair can be performed efficiently.

o Small sized products

ICs were developed to make conventional circuit smaller.
ICs are very compact, light weight and operated efficiently on small power.

o Improvement of operation stability

Stability against variation in external conditions (temperature or power voltage) is improved, unstable factors due to uneveness of individual parts in the conventional circuit are eliminated and reliability is remarkably improved by reducing the number of parts and solderings.

* TYPES OF IC:

- IC
  - Monolithic IC — The most popular IC which uses silicon system base plates to which transistor diode or resistor pellets are wired. Its features are that mass production is possible, element density is great and it is small and light weight.
  - Hybrid IC — Conventional transistor or monolithic ICs used as semi-conductors. It consists a metal coating on which resistors and small type capacitors are placed. Various circuit elements can be used and ICs with the same constants as conventional elements are possible. Capacitors can be used simply.

* IC EMPLOYED IN THIS UNIT: (Refer to Fig. 10)

The ICs used in this unit are "Monolithic linear IC." Two are used in the pre-amplifier section.
The ICs achieve the same performance as that of a circuit using 3 transistors and 5 resistors as shown in Fig. 10.

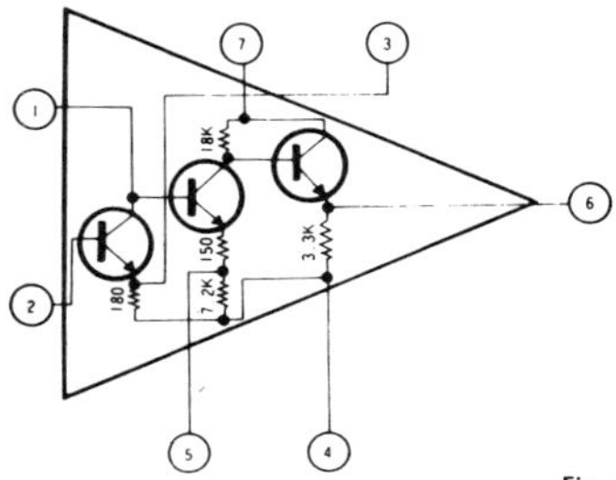

Fig. 10.

* PRECAUTIONS CONCERNING IC:

1. Since the distance between IC terminals is very short, perform measurement of the terminal voltages carefully so as not to cause shorts by contact between the tester bar and the terminal.
2. When checking ICs, stray capacity is induced and abnormal oscillation may occur. Therefore, avoid a long lead line during measurement.
3. When replacing ICs, do not use large capacity soldering, irons. Apply a small capacity iron to the IC terminal for a short time.
4. ICs cannot withstand large voltage and currents. Therefore, exercise care so as not to apply high voltages to IC terminals.
5. Do not bring materials with energized static electricity near ICs. Always ground such materials. IC goes broken.
6. When the soldering iron has or electric potential due an internal failure in the soldering iron, always ground the soldering iron before soldering.

* SIMPLE TROUBLESHOOTING AND REMEDIAL PROCEDURES:

Symptom	Cause and remedy	Figure for trouble location
No power (Unit does not operate the music tape inserted.)	(Cause) Fuse blow (Remedy) Check fuse and replace if defective. (Cause) Imperfect connection of connecting section of power supply lead. (Remedy) Check connecting section of power supply terminal or fuse box.	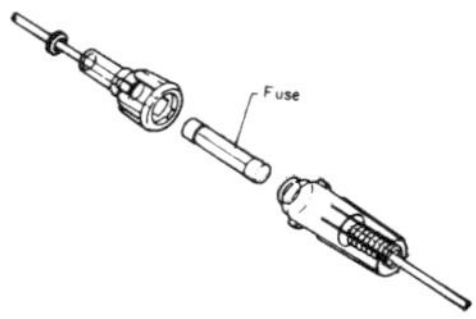
	(Cause) Imperfect grounding of car stereo unit. (Remedy) Ground car stereo unit properly on the car body.	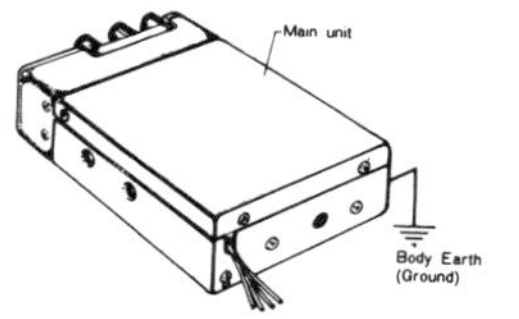
one channel dead	(Cause) Speaker defective (Remedy) Change the speaker connection leads to other terminal. If sound comes out, that channel is defective and if sound does not come out the speaker or speaker connection is improper.	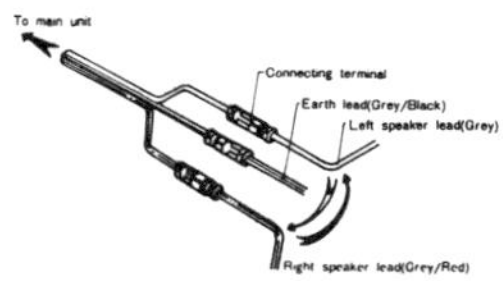
	(Cause) Faulty contact or opening of speaker connection terminal. (Remedy) Check for defects and repair.	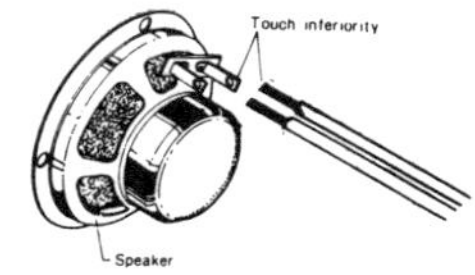
Indicator lights but no sound is produced.	(Cause) Motor does not rotate. (Remedy) Check motor on motor connection (Cause) Shape or recording of tape pack is faulty. (Remedy) Replace tape pack.	
High tone is not produced.	(Cause) Head dirty. (Remedy) Clean the face of head with pure alcohol or head cleaner. (Cause) Faulty adjustment of head azimuth. (Remedy) Readjust head azimuth (Refer to Fig. 9)	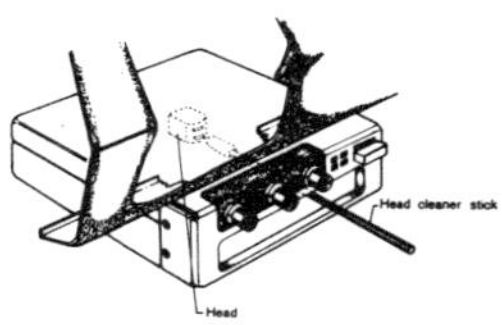

Symptom	Cause and remedy	Figure for trouble location
Wow occurs	(Cause) Tape winding is too tight or pinch roller is deformed or has deteriorated. (Remedy) Replace tape pack.	
Crosstalk occurs	(Cause) Faulty recording of tapes or deformation of tape pack. (Remedy) Replace tape pack. (Cause) Faulty adjustment of head area. (Remedy) Readjust head area. (Refer to Fig. 8) (Cause) Faulty adjustment of head azimuth. (Remedy) Readjust head azimuth. (Refer to Fig. 9)	
Automatical switching channel not possible	(Cause) Sensing tape is defective or solenoid contact dirty. (Remedy) Replace tape pack or clean solenoid contact face with pure alcohol or head cleaner.	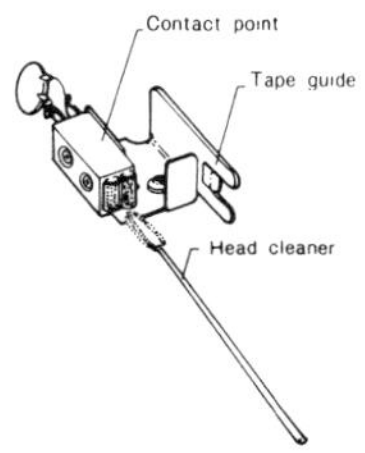

TROUBLESHOOTING OR REMEDIAL PROCEDURES: (mechanism section) Refer to Disassembly Figure.

Symptom	Defective circuit (point)	Trouble location and cause	Remedy
Head switching in possible	Plunger circuit Head switching mechanism	o Faulty contact or defective plunger switch mechanism. o 1S1943 (diode) shorted. o Plunger coil layer shorted. o Solenoid contact defective. o Faulty clutching of ratchet and ratchet kick. o Incomplete wiring of head lead. o Tape defective.	Replace. Replace. Replace. Replace Repair or replace. Repair. Replace.
Crosstalk occurs	Head shift mechanism	o Shape of tape pack improper. o Faulty adjustment of area.	Replace. Adjust.
High or wow abnormal speed.	Driving mechanism system	o Flywheel defective. o Belt defective o Grease accumulated on pulley, belt or flywheel. o Grease or dirt accumulated on capstan. o Oilless bearing broken. o Motor defective. o Tape pack defective. o Tape protector contacts with capstan due to deformation.	Replace (always insert special washer). Replace. Clean. Clean. Replace. Replace. Replace. Repair.
Channel indicator.		o Pilot lamp or lead wire open or faulty soldering. o Faulty contact of rotary switch or accumulated dirt.	Replace or repair. Repair or clean.

* EXPLODED VIEW (See Parts List)

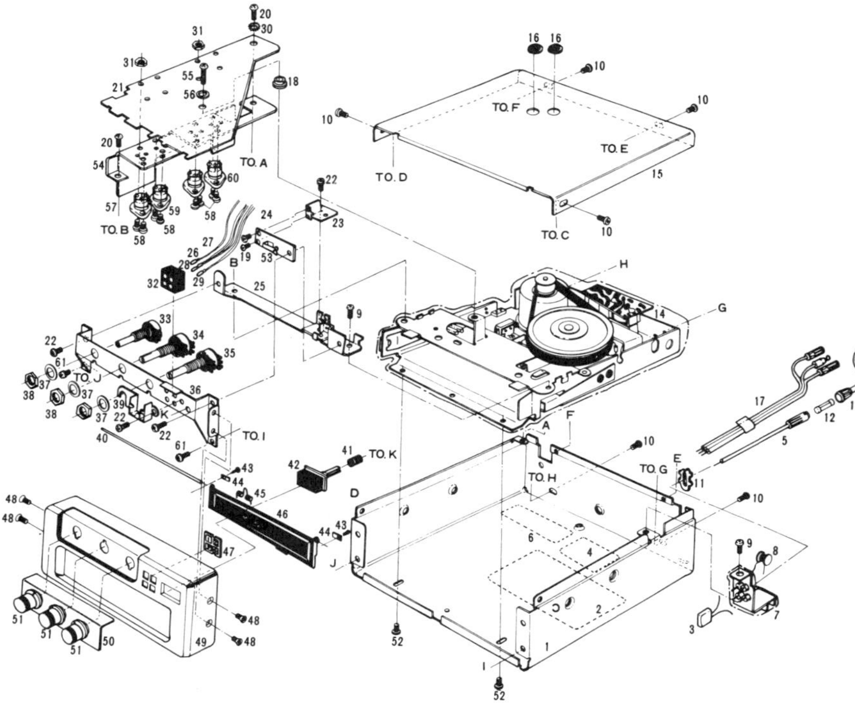

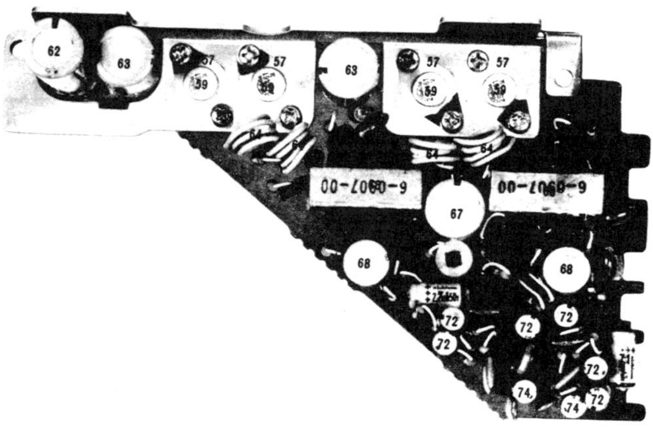

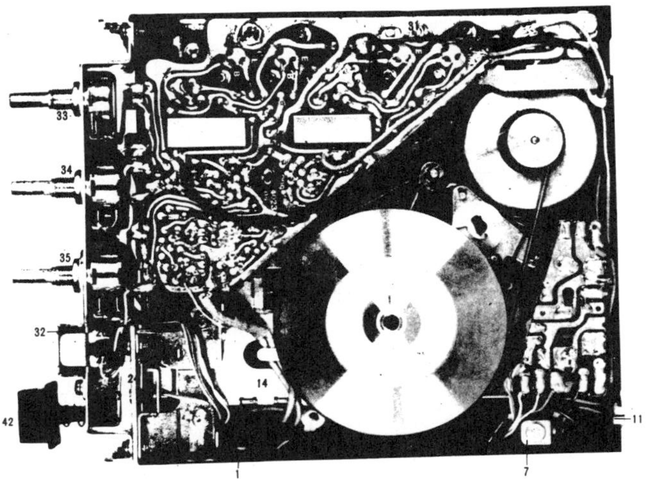

∗ PARTS LIST: (Refer to disassembly diagram. Grey portions in the parts list designate assemblies.)

REF NO	PART NO	DESCRIPTION	P.C.S	REF NO	PART NO	DESCRIPTION	P.C.S
1	311-0802-02	Lower case	1	26	017-0292-07	Pilot lamp (Brown lead)	1
2	285-0545-00	Guide label	1	27	017-0292-08	Pilot lamp (Red lead)	1
3	141-1043-15	Polyester capacitor (50V0.1μF)	1	28	017-0292-09	Pilot lamp (Orange lead)	1
4	285-0583-00	Guide label	1	29	017-0292-10	Pilot lamp (Yellow lead)	1
5	850-1822-00	Power supply cord	1	30	742-3000-20	Toothed washer	1
6	286-3727-00	Set plate	1	31	725-0182-00	Plate nut (Hex type)	2
7	344-0387-00	Filter assembly	1	32	345-2403-00	Pilot lamp mtg rubber	1
8	010-0820-01	Filter coil	1	33	012-3274-00	Variable resistor (BALANCE CONTROL)	1
9	731-3006-80	Tap tight (M3×0.5×6)	2	34	012-3273-00	Variable resistor (TONE CONTROL)	1
10	731-3006-89	Tap tight (M3×0.5×6)	6	35	012-3272-00	Variable resistor (VOLUME CONTROL)	1
11	335-0580-00	Cord clamp	1	36	330-5020-00	Variable resistor mtg plate	1
12	120-0030-00	Fuse (3A)	1	37	745-0485-00	Special washer	3
13	850-1844-01	Power supply cord	1	38	722-0282-00	Special nut (Hex type)	3
14	330-0405-20	Tape mechanism assembly	1	39	330-5045-00	Pilot lamp mtg rubber use plate	1
15	310-0743-01	Upper case	1	40	341-0590-00	Shaft	1
16	335-0618-00	Hole cap	2	41	750-1474-00	Spring	1
17	851-2057-01	Speaker cord	1	42	380-3369-00	Program control button	1
18	335-0622-00	Rubber ring	1	43	704-3008-11	Tapping screw (M3×8)	2
19	714-3006-81	Machine screw (M3×0.5×6)	2	44	330-3030-00	Shaft clamp	2
20	730-3008-80	Tap tight (M3×0.5×8)	2	45	750-1236-00	Spring	1
21	099-4023-06	Printed circuit board	1	46	320-0157-04	Dustproof cover	1
22	732-3006-11	Machine screw sems type (M3×0.5×6)	4	47	375-0506-00	Pilot lamp accessory	1
23	330-5031-00	Program switch mtg plate	1	48	714-3006-36	Machine screw (M3×0.5×6)	4
24	099-4095-00	Printed circuit board	1	49	370-2839-00	Escutcheon	1
25	330-4866-03	Printed circuit board attaching plate	1	50	371-2379-00	Trim plate	1

REF. NO.	PART NO.	DESCRIPTION	P.C.S	REF. NO.	PART NO.	DESCRIPTION	P.C.S
51	380-3368-00	Knob	3	73	051-0011-00 ~04	Monolithic integrated circuit (TA7063P)	2
52	714-3004-89	Machine screw (M3×0.5×10)	2	74	170-4754-32	Electrolytic capacitor (VL16V4.7μF)	2
53	330-5006-00	Program switch plate	1	C1,51,	141-1023-11	Polyester capacitor (50V0.001μF)	2
54	313-0817-01	Heat sink	1	C4,54	144-1021-14	Mica capacitor (50V100PF)	2
55	714-3010-11	Machine screw (M3×0.5×10)	1	C6,56	141-2233-13	Polyester capacitor (50V0.022μF)	2
56	746-0030-00	Fiber washer	1	C8,58	141-3333-14	Polyester capacitor (50V0.033μF)	2
57	079-0001-00	Transistor holder	4	C10,60	141-5123-12	Polyester capacitor (50V0.0051μF)	2
58	714-3008-11	Machine screw (M3×0.5×8)	8	C12,62	141-5133-14	Polyester capacitor (50V0.051μF)	2
59	2SB461	Power transistor	4	C101	141-4733-14	Polyester capacitor (50V0.047μF)	1
60	078-0011-00	Insulating paper	4	R1,51	111-3932-32	Film resistor (¼W 39KΩ)	2
61	731-3006-11	Tap tight (M3×0.5×6)	2	R2,52	111-1032-32	Film resistor (¼W 10KΩ)	2
62	042-0150-00	Electrolytic capacitor (VL16V1000μF)	1	R3,53 5,55	111-1542-32	Film resistor (¼W 150KΩ)	4
63	170-4774-22	Electrolytic capacitor (VL10V470μF)	2	R4,54	111-3322-32	Film resistor (¼W 3.3KΩ)	2
64	031-0010-00	Resistor wire (0.5Ω)	4	R6,56	111-6822-31	Film resistor (¼W 6.8KΩ)	2
65	002-0140-00	Thermistor	4	R7,57 11,61	111-1012-32	Film resistor (¼W 100Ω)	4
66	006-0907-00	Input transformer	2	R8,58	111-2232-32	Film resistor (¼W 22KΩ)	2
67	042-0150-00	Electrolytic capacitor (VL10V330μF)	1	R9,59	111-1042-32	Film resistor (¼W 100KΩ)	2
68	170-1074-22	Electrolytic capacitor (VL10V100μF)	2	R10,60	111-2212-32	Film resistor (¼W 220Ω)	2
69	2SC735	Silicon transistor	2	R12,62	111-4732-32	Film resistor (¼W 47KΩ)	2
70	2SC373	Silicon transistor	2	R14,64 16,66	111-1502-32	Film resistor (¼W 15Ω)	4
71	170-2254-62	Electrolytic capacitor (VL50V2.2μF)	2	R101	111-1022-32	Film resistor (¼W 1KΩ)	1
72	170-1064-22	Electrolytic capacitor (VL10V10μF)	6	R13,63 15,65	110-2211-41	Solid resistor (½W 220Ω)	4

* MECHANISM EXPLODED VIEW (See Parts List)

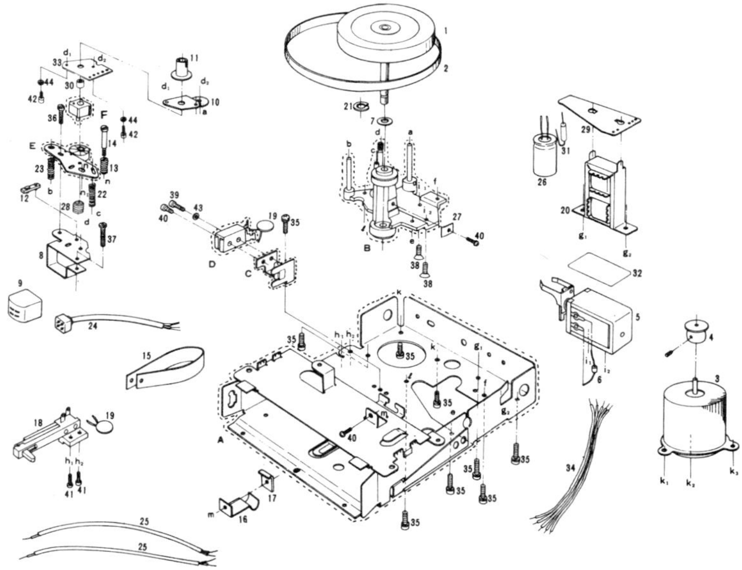

* MECHANISM PARTS LIST:
(Refer to the mechanism exploded view.)

REF NO	PART NO.	DESCRIPTION	P.C.S
1	611-0020-01	Fly wheel	1
2	602-0019-00	Belt	1
3	020-0294-01	Motor	1
4	603-0022-00	Motor pulley	1
5	015-0190-02	Plunger	1
6	001-0077-00	Diode	1
7	746-0620-00	Special washer	1
8	630-0546-00	Head holder	1
9	011-0213-00	Head	1
10	630-0547-00	Rotary plate	1
11	631-0089-00	Area adjustment nut	1
12	631-0090-00	Spacer	1
13	750-1403-00	Spring	1
14	716-0212-01	Special screw	1
15	631-0076-00	Guide plate	1
16	630-0519-00	Spring	1
17	630-0518-00	Spring mtg plate	1
18	013-3059-00	Switch	1
19	043-0022-00	Special capacitor (250V0.04μF)	2
20	009-0585-00	Choke	1
21	743-4000-00	E-Ring (M4)	1
22	750-1410-00	Spring	1
23	750-1411-00	Spring	1
24	852-4559-01	Extension lead	1
25	815-0890-02	Shielded wire	2
26	042-0152-00	Special capacitor (VL16V470μF)	1
27	630-0612-00	Tape protector	1
28	750-1475-00	Spring	1
29	099-4041-03	Printed circuit board	1
30	632-0302-00	Sleeve	1
31	115-1212-61	Film resistor (2W120Ω)	1
32	347-0339-00	Insulating paper	1
33	099-4042-02	Printed circuit board	1
34	816-0089-00	Special lead	1
35	732-3006-11	Sems screw (M3$\times$6)	9
36	715-3010-61	Machine screw (M3$\times$10)	1
37	715-2610-61	Machine screw (M2.6$\times$10)	1
38	714-3006-41	Machine screw (M3$\times$6)	2
39	714-2608-11	Machine screw (M2.6$\times$8)	1
40	714-2606-11	Machine screw (M2.6$\times$6)	3
41	714-2308-11	Machine screw (M2.3$\times$8)	2
42	714-2006-11	Machine screw (M2$\times$6)	2
43	711-2600-21	Spring washer (M2.6)	1
44	711-2000-21	Spring washer (M2)	2

* MECHANISMS ASSEMBLIES PARTS LIST

REF NO	PART NO	DESCRIPTION	P.C.S
A	960-2707-00	Deck plate assembly	1
	600-0041-01	Deck plate	1
	606-0036-01	Pack guide	1
	960-2627-00	Roller assembly	1
	728-3040-87	Rivet	1
B	960-2733-04	Housing assembly	1
	607-0016-01	Housing	1
	609-0027-00	Bearing	1
	609-0028-00	Bearing	1
	612-0100-04	Shaft	1
	612-0098-01	Shaft	3
C	960-2705-01	Tape guide assembly	1
	605-0026-00	Tape guide	1
	631-0109-00	Cap	1
D	960-0360-02	Contact assembly	1
E	960-2734-00	Cam assembly	1
	608-0029-02	Cam	1
	630-0548-00	Stay	1
F	960-2732-00	Ratchet assembly	1
	631-0075-01	Ratchet	1
	750-1401-01	Spring	1

〔Instrument Check〕

If defective points can not be localized by means of the tester or other methods, check with instruments.

1. Instruments employed Low frequency osillator, Osilloscope. Millivalve voltmeter.
2. Measurement method Detect defective circuits by connecting the live terminal of the osilloscope to check points CP_5, CP_7 and CP_{15} in that CP_4. (Check the left channel in the same manner as the right channel.)

(Trouble example)

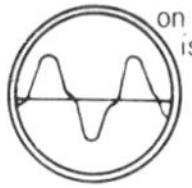

If the waveform shown (crossover distortion waveform) on the left 15 appears at CP_{15}(R) the output circuit is defective.

{ Thermistor
{ R_{63}, R_{64}, R_{65}, R_{66}

If the waveform on the left appears CP_5(R), the pre-amplifier circuit is defective. (R_{51}, R_{52}).

If the waveform shown on the left appears at CP_7(R), the drive circuit is defective. (R_{58})

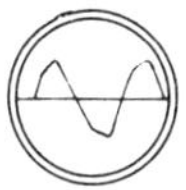

If the waveform shown on the left appears at CP_{15}(R), the drive circuit is defective. ($R_{58, 59, 61}$)

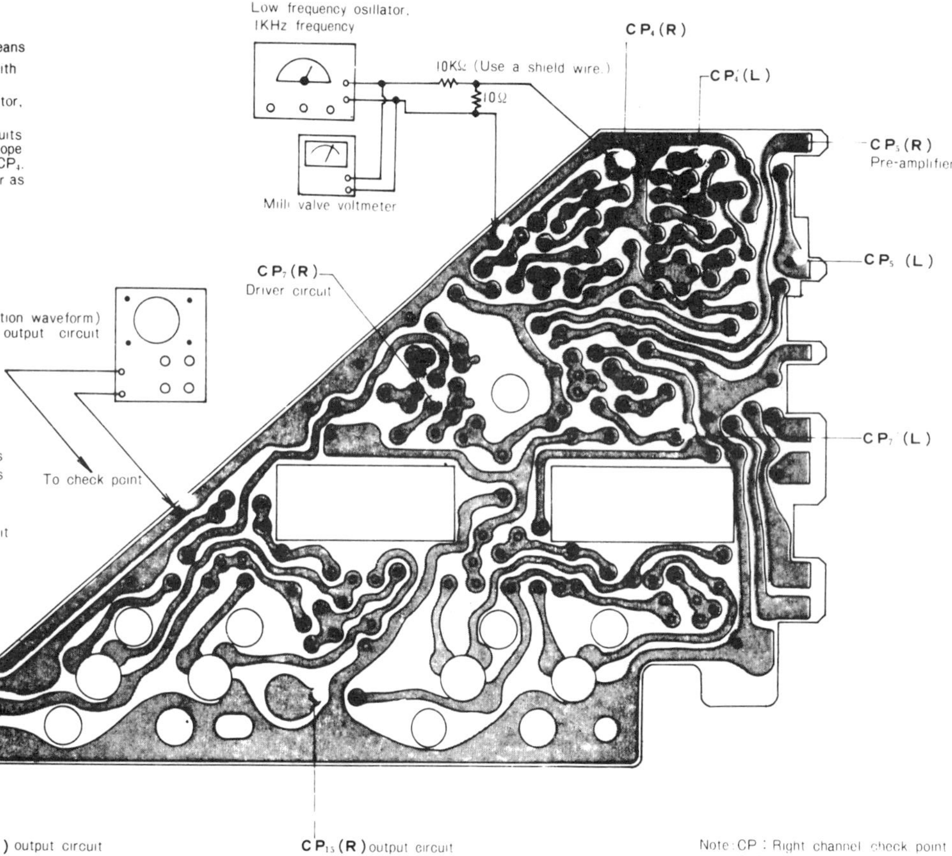

Note: CP : Right channel check point
CP' : Left channel check point

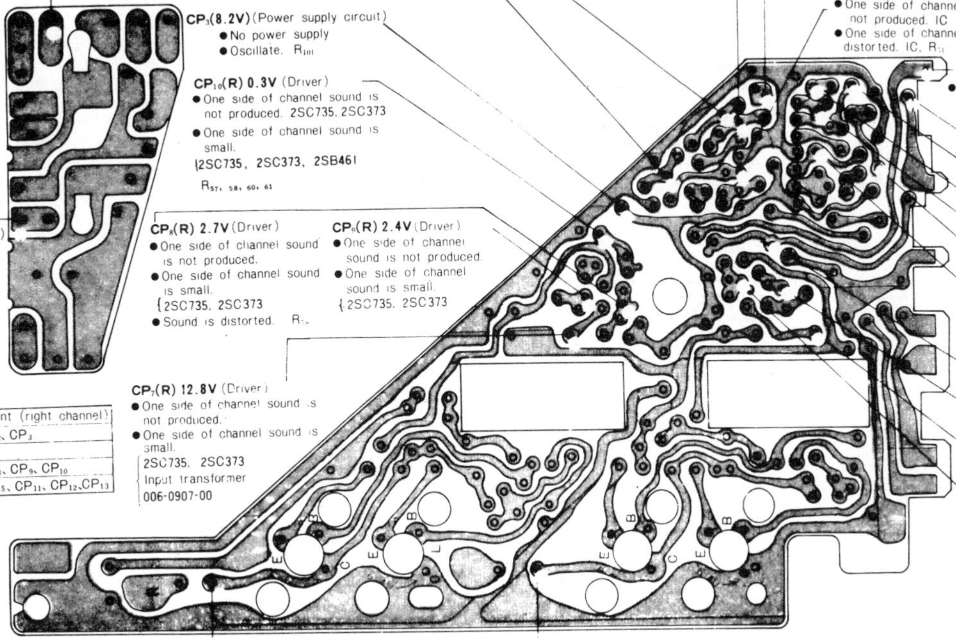

Circuit	Check point (right channel)
Power supply circuit	CP$_1$, CP$_2$, CP$_3$
Output circuit	CP$_6$
Drive circuit	CP$_7$, CP$_8$, CP$_9$, CP$_{10}$
Pre-amplifier circuit	CP$_4$, CP$_5$, CP$_{11}$, CP$_{12}$, CP$_{13}$

(Check the above)
When both channels are defective-From check points in power supply circuit.
When one side channel is defective-From check points in each circuit.

Note : CP : Check points for right channel.
CP : Check points for left channel.
Described voltage : Normal voltage between CP and ground at no signal.
Testers employed : Internal resistance 4KΩ/V
10V or greater 50V range
2.5V or greater 10V range
Below 2.5V 2.5V range.

* PRINTED CIRCUIT BOARD

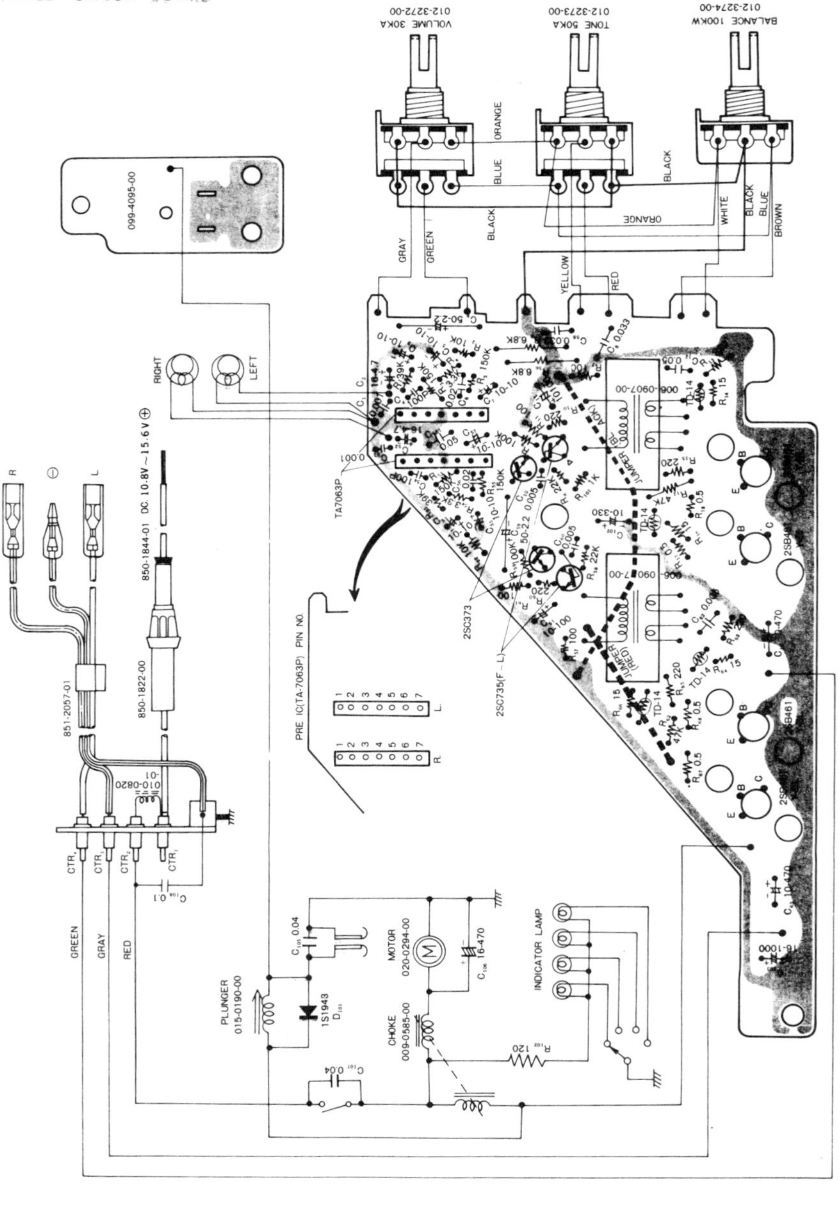

* **CIRCUIT DIAGRAM:** (Voltage and current values entered.)

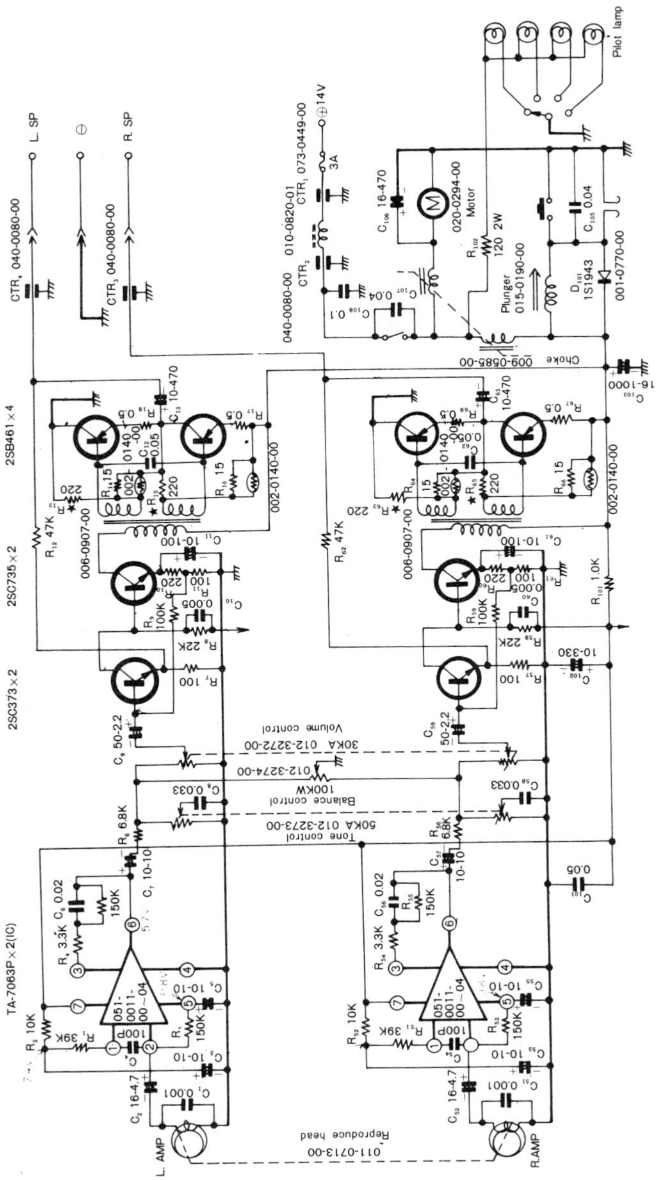

NOTE 1. Unless otherwise carbon film resistor $\frac{1}{4}$W ±10%

NOTE 2. Resistor marked with ★ are for carbon resistor $\frac{1}{2}$W ±5%

NOTE 3. Conduct measurements. Under no signal conditions with a tester having an internal resistance of 4KΩ/V and at the following ranges:

2.5V or greater 10V range
0.5V or greater 10V range
Below 0.5V 0.5V range

Measurement value may be different, if measured by wrong range.

NOTE 4. Fuse of 3A.

Chapter 7

Servicing and Replacing Recording Components

Repairing the recording section of a tape recorder is not difficult once you understand the basic theory of magnetic recording.

Recording heads and the way they function

All recording heads are basically similar. Most inexpensive machines use combination heads (the same head is used for both recording and playback) while more expensive equipment uses separate recording and playback heads.

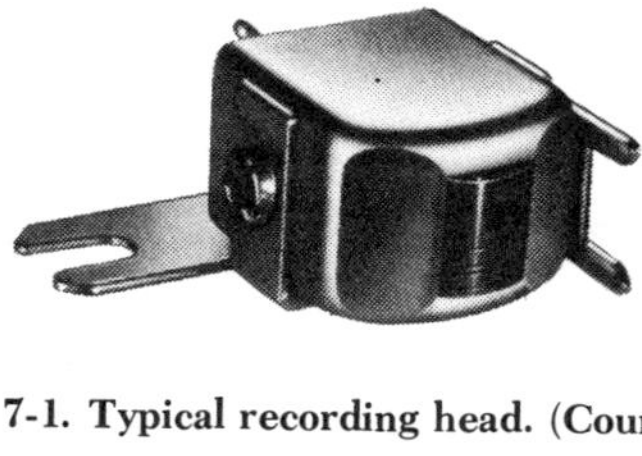

Figure 7-1. Typical recording head. (Courtesy of 3M)

The head is an electromagnet which is placed against the oxide coating on the tape. Current flows through the coil and magnetic lines of flux flow through the oxide (across the narrow gap in the tape head) and the iron oxide is thus magnetized in varying patterns. Audio signals are stored on the tape according to the strength and polarity determined by the amplifier, which has sent the signals it has received from the microphone (which has converted sound waves into electric impulses).

To be precise, however, the process is not quite that simple. Since the relationship between the current from the head and the magnetic fields are not "linear," distortion results. By using *ac bias* (a high-frequency—30 to 100 kHz—sine wave) fed into the tape head along with the signal which is recorded, this distortion is eliminated.

Checkpoints for instant analysis of recording malfunctions

In many cases, the reason for equipment not operating in the record mode is mechanical, not electronic. For example, cassette tapes have interlock tabs; if they are removed, the record lock lever in the cassette recorder will keep the machine from recording (as a safety device to prevent accidental erasure). Also, some pieces of equipment use microphones with a switch that turns the recorder on and off. (Customers sometimes forget about this switch.) In addition, dirty, damaged or worn heads will cause recording trouble.

Figure 7-2. Cassette interlock tabs can be removed and replaced with tape for re-recording. (Missing tabs keep equipment from recording.) (Courtesy of 3M)

Test the machine carefully. Put a blank tape in the machine and try to record. Next, double check to see if the record mode is inoperable by playing a prerecorded tape (if it doesn't play, you may have other problems). In tracing the trouble, use the following list to analyze the manfunction in seconds.

1. *If it won't record*
   a. use another tape. (Replace missing interlock tabs in cassettes with tape.)

b. clean the recording head.
c. check for an open jack at the mike.
d. demagnetize heads.
e. check to see pressure pads keep tape against head.
f. replace the defective speaker or earphone.
g. replace the defective record switch.
h. repair the defective bias oscillator.
i. check and correct the power supply voltages (usually no B+ or low B+).
j. correct defects in the amplifier (replace defective transistors, capacitors or resistors).
k. replace the defective head.

2. *If the sound is distorted*
   a. correct the low bias oscillator output.
   b. replace the defective power supply component.
   c. replace defective transistors, capacitors and resistors in recording circuit of the amplifier.
   d. replace the tape with a high quality tape.
   e. align or replace the faulty VU meter.
   f. adjust the recording bias.
   g. adjust the recording current.

3. *If high frequency sounds are missing or are weak*
   a. clean and adjust the head.
   b. align the head (azimuth adjustment).
   c. check the polarity of the leads to the head.

4. *If hum is present in the recorded signal*
   a. replace the unshielded leads at the head (or keep them from touching or shorting).
   b. replace the defective or unshielded cord at the microphone.

Troubleshooting the recording section of conventional and Dolbyized units

The same amplifier is used for both recording and playback in most tape equipment. To track down and correct malfunctions, use the same techniques that you would apply with any electronic equipment.

The conventional recording process is electronically simple. In *record* mode, the signal from the MIC input is applied to the base of one transistor, amplified, applied to the bases of other transistors (and amplified further), and then divided (part of the signal goes to the record level control), with part of the signal going

MAKING AN ORDINARY RECORDING

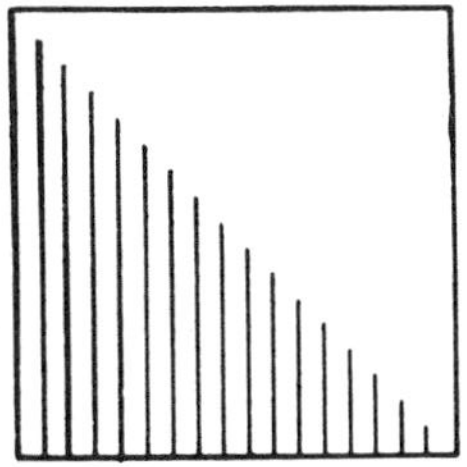

1. Music. Music is made of sounds of different loudness separated by intervals of silence. Loud and soft sounds are here shown as long and short lines. The music represented by this diagram starts loud and gradually becomes very quiet.

2. Noise. Any recording tape, even the best kind, makes a constant hissing noise when played. At the very slow speeds and narrow track widths used in tape cassettes, tape noise is much more noticeable than it is in professional tape recordings, although even there it is a problem.

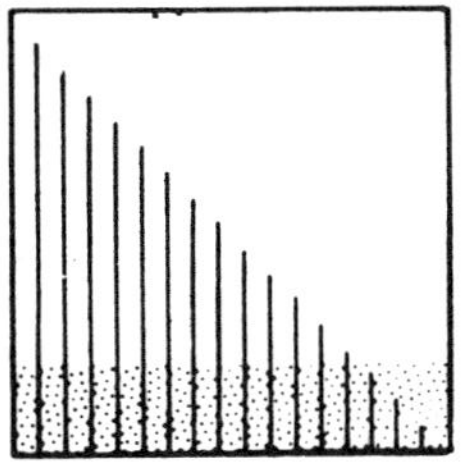

3. Music and Noise. When a tape recording is played, the noise of the tape conceals the quietest musical sounds and fills the silence when no sound should be heard at all. Only when the music is loud is the noise not usually heard; however, tape noise is so different from musical sounds that it sometimes can be heard even then.

MAKING A DOLBY SYSTEM RECORDING

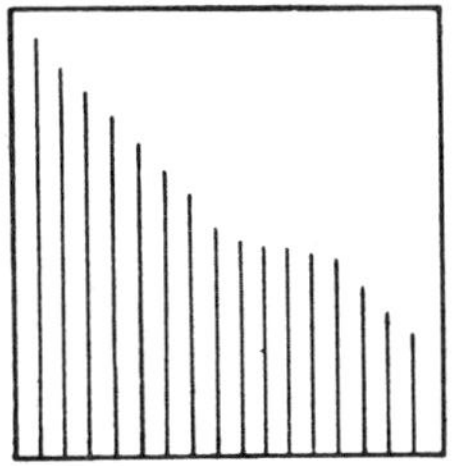

1. What the Dolby System does first. Before the recording is made, the Dolby System "listens" to the music to find the places where a listener might later be able to hear the noise of the tape. This happens mainly during the quietest parts of the music. When it finds such a place, the system automatically increases the volume of the higher frequencies so that the music is recorded louder than it would be normally.

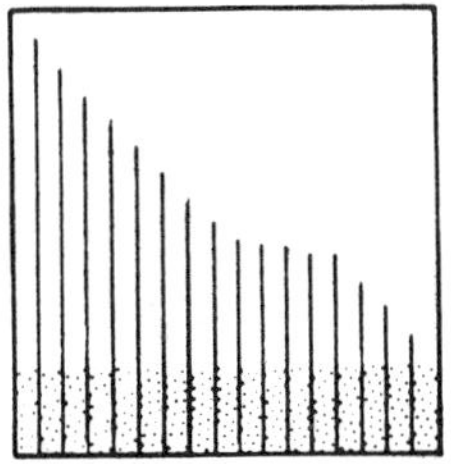

2. The Recording. In a Dolby System recording, the parts of the music which have been made louder stand out clearly from the noise. As a result, Dolby System recordings sound brilliant and unusually clear even when played back without the special Dolby System circuit.

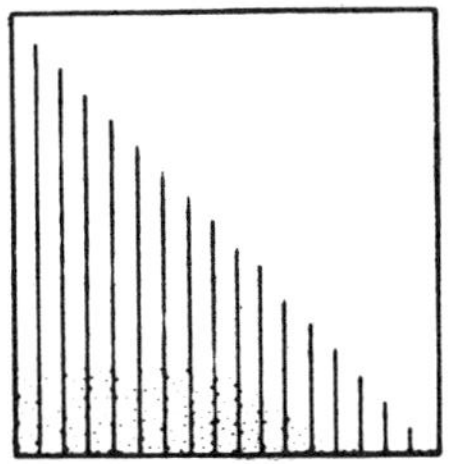

3. What the Dolby System does during playback. When the tapes are played on a high-fidelity tape recorder equipped with the Dolby System circuit, the loudness of the higher frequencies is automatically reduced in all of the places at which it was increased during recording. This restores the music to its original fidelity again. At the same time, the noise which has been mixed with the music is reduced in loudness by the same amount — usually enough to make it inaudible.

Figure 7-3. What the Dolby system does. (Courtesy of 3M)

to another transistor base, again being amplified and then coupled to the head through a driver.

Dolbyized units or Automatic Noise Reduction Systems (ANRS) are designed to reduce noise introduced during recording: hiss, bias noise. These systems also reduce noise during playback. This is done by increasing the amplitude of certain frequencies.

Since noise is introduced during recording, the Dolby process takes place before the amplifier stage. The signal is passed through eight stages: through the attenuator, through the emitter follower, and through the variable high pass filter where the signal is divided in two. Of these two signals, one is sent directly to the amplifier, and the other (with low level frequencies) is sent to a high gain amplifier.

The Dolby system (a patented circuit that "listens to music" and adjusts the recording process accordingly) is really no more than a separate circuit (containing easily serviced components) in a conventional amplifier circuit.

Adjusting Dolbyized equipment requires schematics and manufacturers' specifications. For example, to adjust the Dolby record circuit on one popular model, you will need an audio frequency oscillator and an attenuator (600 ohms) sending in a 5 kHz, −10 dB signal with a VTVM attached to the output. Make a reading with the Dolby switch *on* and a reading with it *off*. There should be a difference of 8 dB (plus or minus 0.5 dB). The level difference between channels should be less than 0.5 dB. To make a correction, adjust components by turning a screw in or out.

Obviously, to save servicing time, you need to get a schematic of the recorder that you are working on. The specifications listed will not fill the requirements of every Dolby unit in every recorder. If you are specializing in the products of one manufacturer, you will find that you will be able to make such adjustments almost automatically. Otherwise, you will have to exercise patience and search the specifications.

Replacing worn recording heads

Worn heads cause a loss in high frequency response. The iron oxide coating on the tape moving past the heads works like sandpaper to gradually wear the heads down. For a time the response is unaffected (in fact, it may even be better when the head becomes

The principles of ANRS

One characteristic of the human ear is its "masking effect". When two sounds occur at the same time, and there is a big difference in level, only the louder sound will be heard. ANRS reduces tape hiss which occurs in the middle and high frequency ranges, and amp noise and bias noise during recording. It also reduces amp noise during playback. ANRS does not alter the music signal in any way.

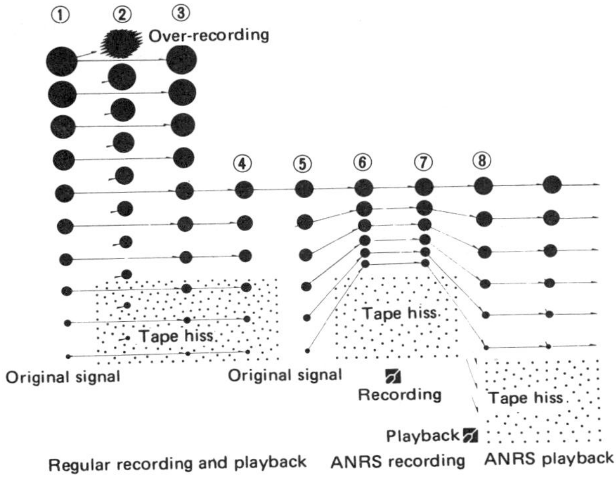

The illustration will help you understand ANRS.

(1) The black circles represent sound; the bigger the higher the level, and the dots represent noise caused by tape hiss.

(2) When the recording level is increased too much the tape will become saturated and the recorded signal distorted.

(3) For good recordings it is essential that the recording level meter should not exceed 0VU. So the low level signals overlap with tape hiss, a low level noise inherent in tapes. When a loud passage is played the tape hiss is not noticeable.

(4) But when the loud passage is followed suddenly by a quiet passage annoying tape hiss can be heard.

(5) And this is when ANRS starts to work.

(6) Low level signals alone are raised in level by an amount which is determined by their level and frequency. The level to which they are raised is above the level of the tape hiss and so tape hiss and music are not recorded at the same level.

(7) The recorded signal contains both music which has been modified by ANRS and tape hiss at a lower level. The music and tape hiss are at different levels. Because the music has been modified by ANRS it cannot be played back directly.

(8) In playback ANRS works in exactly the opposite way to which it works in recording. It performs a complementary operation, and restores transformed music signal to the original sound. At the same time the level of the tape hiss is also reduced, making it inaudible.

Figure 7-4. ANRS (Courtesy of JVC)

smoother); then additional wear makes the gap wider and response falls off. At that point the head must be replaced.

In addition, some heads will prove to be defective for a variety of reasons. You can test a head with a signal tracer, a signal injector, or an ohmmeter. Connect the signal tracer to the head coil to see if you can pick up a signal while the tape is running through the machine. If you can, and no sound comes from the amplifier, the trouble is in the amplifier.

An alternate test is to send a signal from a point in the circuit after the coil in the head with the signal injector. If the signal comes through, the trouble is in the head.

Still another method is to connect an ohmmeter across the unsoldered leads. Resistance should measure between 250 and 500 ohms.

Unplug or unsolder the malfunctioning or worn head (after drawing a sketch of the wiring connecting it) and replace it. Unfortunately, not all heads have readily available replacements. Several manufacturers supply books listing interchangeable replacement heads for many machines.

Cost is a decisive factor in head replacement. In many cases, you 'll find the heads (and labor charge) will be more than half the value of the equipment. In such cases, you might suggest that the customer would do better to get a new machine. Give him a good price on the spot and chances are you'll be making profits from a quick and easy sale.

Cleaning and aligning recording heads through a step-by-step process

Recording heads accumulate oxide particles which can be removed with a can of tape head spray cleaner (which comes with a five-inch extension tube to make the job easier), a soft cotton swab dipped in denatured alcohol (or commercially available tape cleaner), or prepackaged open-reel, cassette, or cartridge cleaners (which are run through the machine like any regular tape).

Avoid getting cleaner on any plastic parts and be sure that the head is dry before running tape through the machine. (These problems are avoided, of course, in using the self-contained cleaner tapes.) Avoid scratching the polished surface of the head or damaging any leads. A good cleaning is usually needed every six months.

Figure 7-5. Check the head for wear.

Misaligned heads cause distorted sound and damaged tape. Head alignment varies according to the type of recorder. Follow these simple instructions:

Open-reel recorders are easiest to work on. After the head is mounted, adjust the height. Center the tape between the tape guides or position it with the top edge of the tape head core visible above the tape. Next, apply a 15 kHz signal to the MIC jack at a level which will produce −20 dB output. Put the recorder into *record* mode and adjust the *azimuth* for maximum output. (For combination heads, you can use a commercial test tape to adjust for correct playback and the *record* position will also be correct.) Use a blank tape to make a test recording.

Cassette players are more compact and are thus more difficult to work on because the space is cramped. The basic procedure is the same. Mount the head. Adjust the height so that the tape edge is a hair's breadth below the top of the tape core edge (use a special cassette with the edge cut out to make this positioning easier.) Introduce a 15 kHz signal at the MIC input (−20 dB) and adjust the *azimuth* for maximum output. Test with a blank tape.

Figure 7-6. How to adjust the azimuth.

Figure 7-7. How to adjust the height.

Cartridge players require careful height adjustment as described in Chapter 6. After setting the balance at maximum for the left channel, adjust the screw for a NULL reading on the oscilloscope (connected across the left channel). Adjust the *azimuth* (with the *balance* turned to maximum for the right channel) by turning the screw for a MAXIMUM reading on the oscilloscope (attached to the right channel). Make these adjustments several times. Cartridge machines which record and cassette machines can also be adjusted by using a playback tape if a single head is used for record and playback.

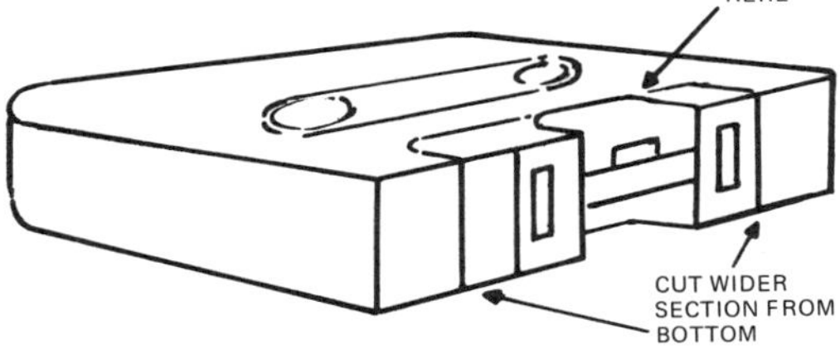

Figure 7-8. Cut away part of a cassette to permit a visual height check.

Fixing other recording components

Electronic repairs usually are easy; tape amplifier circuits are simple. You will need to replace transistors (after testing them in the conventional manner)—or tubes, if you are servicing somewhat older equipment. Resistors and capacitors which have lost their value will also be changed. These are operations which are commonplace.

On occasion, however, you will need to adjust *bias* (which is not adjustable in all equipment). Part of the *bias oscillator's output* is added to the recording signal (the rest is fed to the erase head). Changes in amplitude will cause distortion. To correct this, measure the voltage across the head with a voltmeter (with no signal applied) and adjust according to the manufacturer's specifications (usually about 50 to 100 kHz).

The *bias oscillator wave form* (which should be a perfect sine wave) can be examined with an oscilloscope connected to the head.

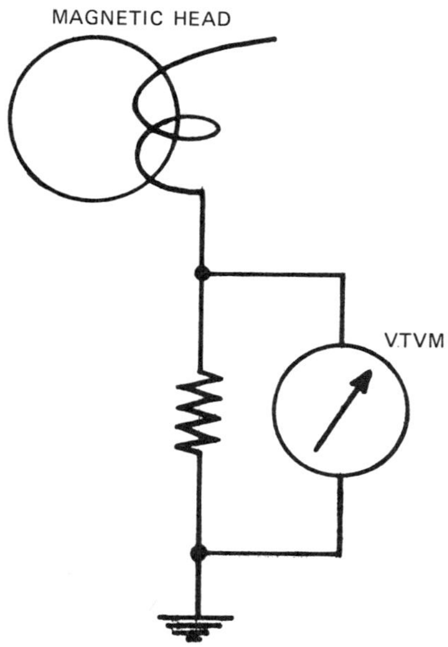

Figure 7-9. How to measure bias current (adjust semi-fixed volume screw to equal 490 μ A—normal tape, 600 μ A chrome tape.)

The *bias trap* can also produce distortion. Adjust it (on the side away from the head) for minimum voltage (with no signal present).

Recording level meters can be adjusted on some machines (on others replacement is necessary). Apply a 1 kHz signal to the MIC jack and adjust it so that the output level of the line-out is 0 dB. Adjust the variable condenser so that the point of the level meter indicates 0 *VU*; then put the equipment into the *playback* mode and adjust the variable condenser so that the playback output reads 0 dB on the level meter.

Adjustments of these components, however, are performed relatively seldom. Most of your servicing will be confined to replacing record and playback heads or servicing the recording section of the amplifier.

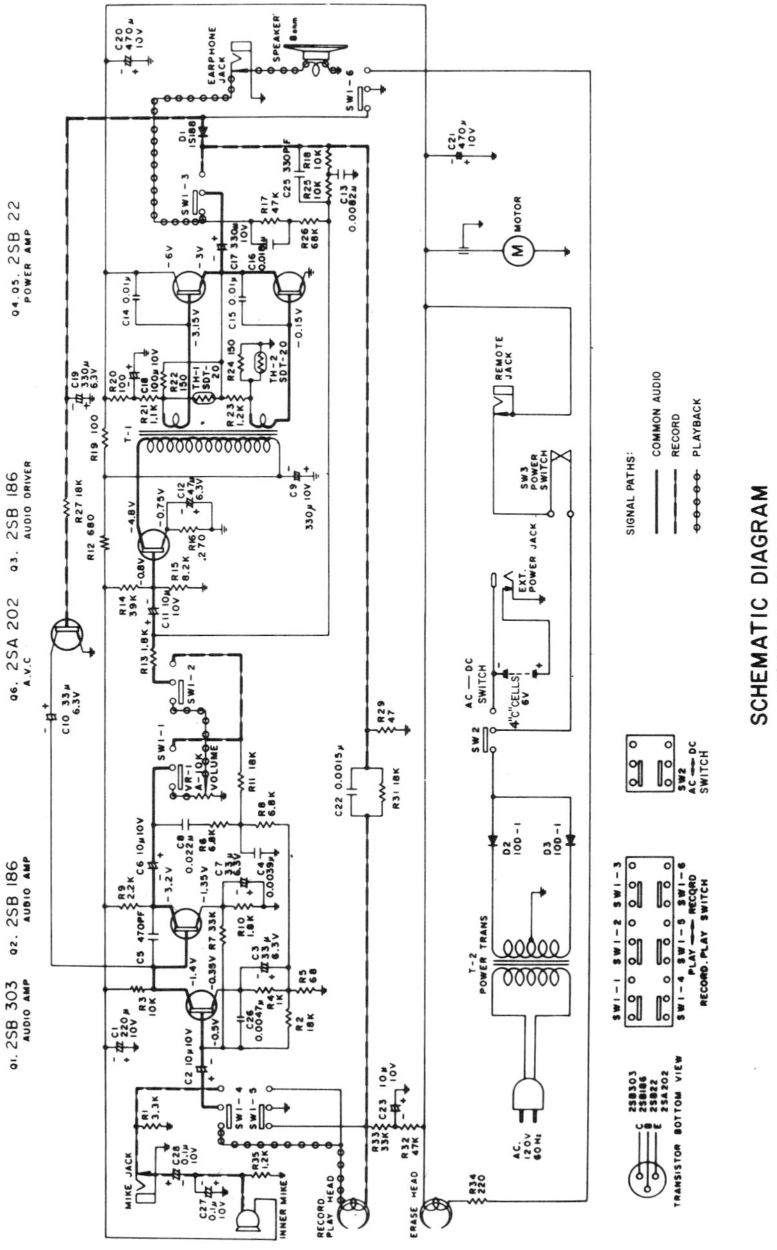

Figure 7-10. Schematic diagram. (Courtesy of General Electric)

Chapter 8

Servicing and Replacing Playback Components

The playback section of a tape recorder/player consists of a playback head and an amplifier.

Playback heads and how they work

Combination heads (recording and playback) are used on many inexpensive machines; quality equipment makes use of separate heads.

Tape is moved past the head (an electromagnet) at the same speed used in recording. Varying magnetic intensities in the tape (created by varying intensities in the electromagnetic signals in the recording process) induce voltage in the head by a simple principle of magnetism: changes in a magnetic field around a coil induce voltage in the coil. The playback head thus cuts through these changing magnetic lines of flux, creating voltage in the head which is then amplified.

This is done through the following stages: the pre-amp (usually a standard direct coupled amplifier), an audio amplifier, a driver amplifier, and an output stage. Tape decks usually have only pre-amplifier stages; therefore, they must be connected to external amplifiers and speakers.

Recorder/players use an equalizer circuit which amplifies high frequencies so that they are not lost during playback. The bias oscillator, necessary in recording, is not used in the playback mode.

BLOCK DIAGRAM

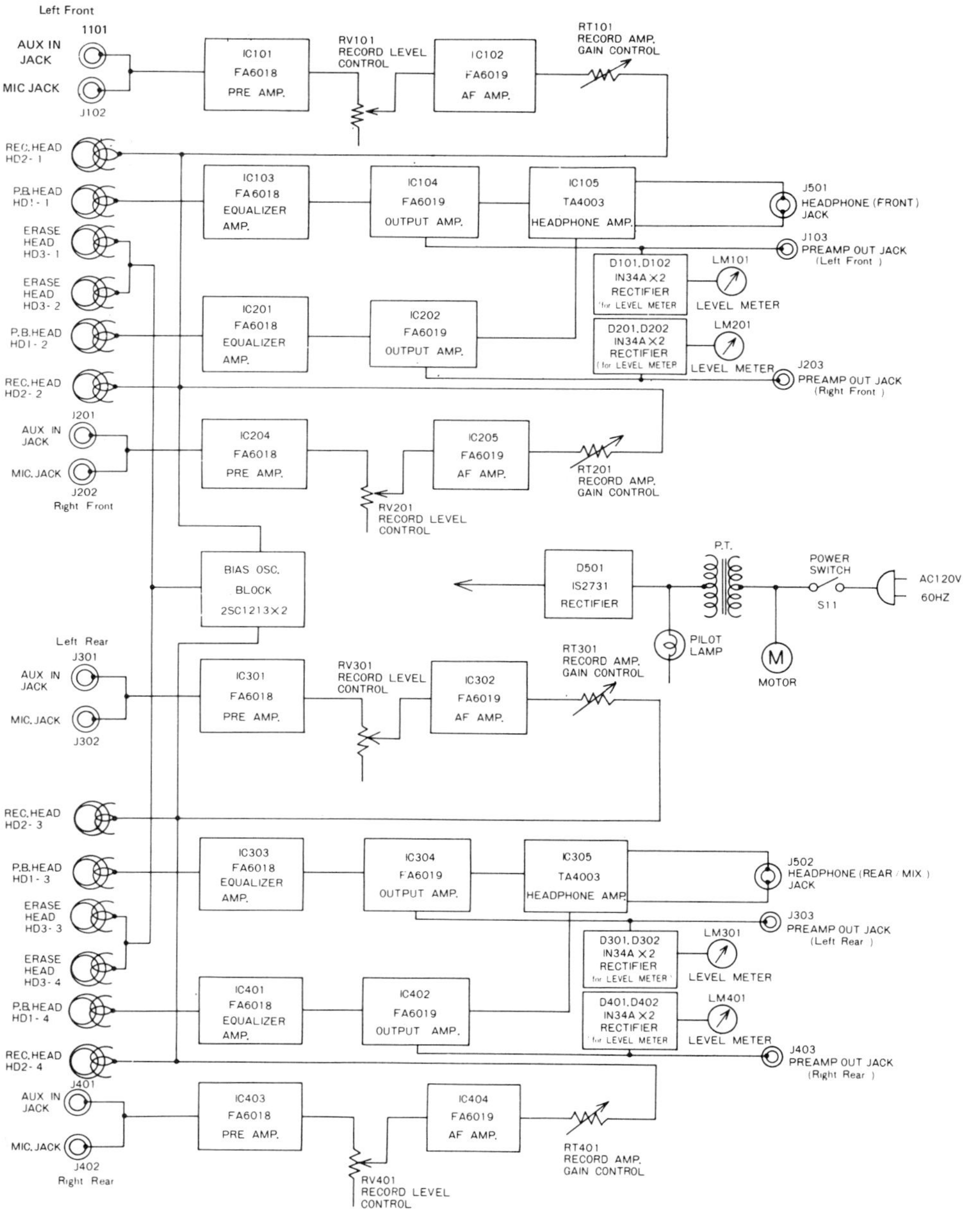

Figure 8-1. A block diagram of an amplifier circuit. (Courtesy of Radio Shack, a Tandy Corporation Co.)

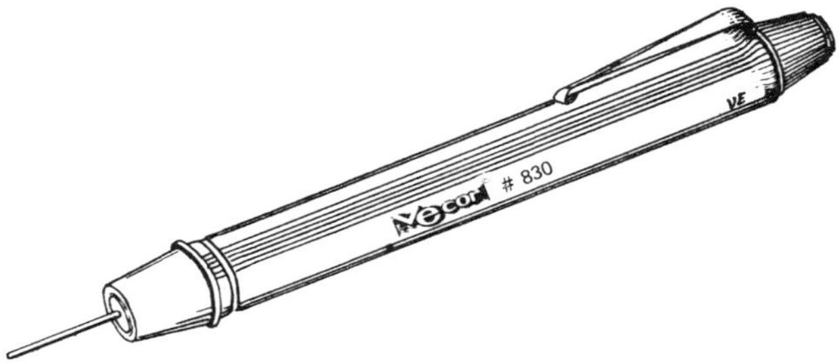

Figure 8-2. Pocket signal tracer. This pocket-sized device generates a 1 KHz signal at .55 volts to trace circuits back to the RF stage. Stoppage of the 1 KHz tone which it emits indicates the problem is in the preceding section. (Courtesy of Vidaire Electronics Manufacturing Corporation)

Ways you can tell if playback heads are worn or damaged

For a quick test of a playback head, connect a signal injector across the playback head. If a signal comes through the amplifier but no sound comes through when you play a tape, you have a defective head.

You can also use an ohmmeter to check whether the coils are shorted or open (disconnect the head from the amplifier to get an accurate reading). Obviously, you can also use a signal tracer to search out the signal from a test tape through the various stages.

Examine the playback head to see if it is loose, scratched or dirty. Use a high-powered magnifying glass or a jeweler's loupe to see if the head is gouged or badly worn. Check to see the tape head is properly aligned.

Since visual inspection has its limits (the head gap specifications of a top quality playback head may be 150 millionths of an inch—impossible to measure), you should listen for limited high frequency response (using a dependable tape and an oscilloscope if possible). Adjust the azimuth and height to specifications before deciding to replace the head.

Pinpointing troubles in playback systems (Dolbyized and conventional)

Use a good quality tape to test the machine. Clean and demagnetize the heads. Then proceed to use this list to find the trouble in one or more areas.

1. *If it won't play*
   a. use another prerecorded tape.
   b. clean the dirty heads.
   c. connect open or shorted leads.
   d. adjust the heads (height, azimuth).
   e. correct the B plus voltage (in battery sets) which is too high or too low.
   f. replace the defective playback head.
   g. check for defective resistors and capacitors in the amplifier circuit.
   h. repair or replace the defective input transformer.
   i. replace the defective volume control (potentiometer).
   j. check for defective or deteriorated transistors (or tubes).
   k. replace faulty power supply components (fuses in automobile installations).

2. *If the sound is weak*
   a. clean and adjust the playback head.
   b. check the pre-amplifier circuit (resistors, capacitors, and transistors or tubes).
   c. replace defective resistors, capacitors or transistors (or tubes) in the amplifier circuit.
   d. check the driver stage output for leaking capacitors, defective resistors or deteriorated transistors.
   e. replace the defective playback head.

3. *If the treble tone is poor*
   a. replace the defective tone level.
   b. replace open capacitors in the inverse feedback circuit.
   c. replace open capacitors in the amplifier circuit.
   d. correct the faulty azimuth of the playback head.

4. *If the bass tone is poor*
   a. correct the faulty azimuth of the playback head.
   b. replace open capacitors in the pre-amplifier circuit.
   c. replace the faulty playback head.
   d. replace shorted or open capacitors in the amplifier circuit.

5. *If the sound is distorted*
   a. correct the power voltage which is either too high or too low.
   b. clean, adjust or replace the playback head.
   c. replace defective capacitors, resistors, or transistors (or tubes) in the driver stage.
   d. check for malfunctioning resistors or capacitors in the power supply circuit.
   e. replace defective transistors, capacitors or resistors in the pre-amplifier or amplifier circuit.

6. *If the output from one channel is low*
   a. adjust the weak pad pressure on the head.
   b. clean, align or replace the playback head.
   c. correct the defective component in the amplifier circuit (capacitor, resistor, or transistor).
   d. match the speaker impedance.

7. *If no sound comes from one channel*
   a. correctly adjust the balance control.
   b. clean and align the playback head.
   c. correct the open or shorted playback head.
   d. check for a broken or shorted connecting cable.
   e. correct the miswired playback head.

8. *If the sound is noisy*
   a. clean or replace the playback head.
   b. replace defective amplifier leads.
   c. replace the defective motor.

9. *If there is a loud hum (with the volume at minimum)*
   a. replace the defective lead or cable.
   b. replace the defective filter.
   c. replace malfunctioning transistors.

10. *If there is a loud hum (with the volume at maximum)*
    a. replace the defective cable or lead.
    b. replace defective filter capacitors.
    c. correct improper or shorted ground leads in amplifier circuit or at head.
    d. replace defective transistors in the driver stage.
    e. demagnetize the head.
    f. replace the defective input stage transistor(s).
    g. align the trap coil.
    h. replace the defective power supply component.
    i. replace the defective motor.

11. *If wow-and-flutter occurs*
    a. See Chapter 4, 5 or 6.

12. *If cross-talk occurs*
    a. replace the defective cartridge or cassette.
    b. reload the incorrectly inserted cartridge in the player.
    c. align height of the playback head.
    d. adjust the tape guide.
    e. correct or replace the defective tape head (which does not move up and down—see chapter on cartridge transport mechanisms).
    f. replace the defective cam (see chapter on cartridge transport mechanisms).
    g. replace the defective head plate spring (see chapter on cartridge transport mechanisms).

How to replace any playback head (techniques for alignment—stereo and quadraphonic)

Once you have identified a head as defective, you must find out if a replacement is available (interchangeable heads are cross-indexed in various catalogs available from your electronics supply house).

Take the time to make a brief sketch of the connections so there is no mistake about what goes where. Unplug or unsolder the old head and replace it with the new one.

You should next align the head. You can do this by ear, but the instruments on your workbench will help you make more accurate adjustments.

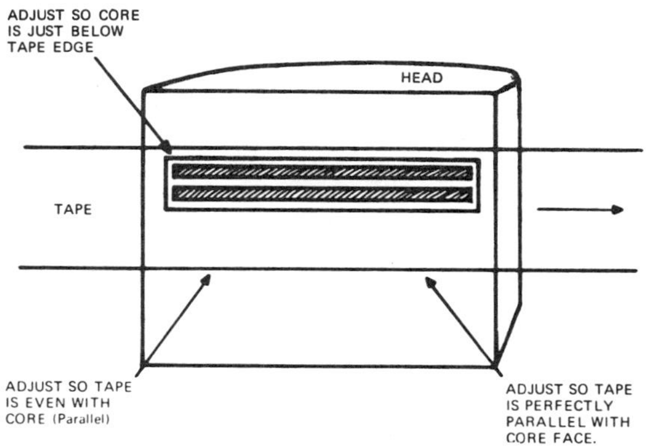

Figure 8-3. How to align heads.

First, align the *height* by adjusting the tape guide with a screwdriver until the top edge of the tape core is visible above the tape; then adjust carefully until the edge of the tape just overlaps the head core.

The *zenith* or angle at which the tape hits against the playback tape head can be checked visually; the tape must be exactly parallel to the tape head.

The *azimuth* is checked by using an alignment tape. Set both the right and left channels to read 0 dB on the VTVM (connected to the output). Play the alignment tape, introducing a 15 kHz signal into the recorder's playback head. Adjust the azimuth adjustment screws until you get a peak reading.

Paint the screws with a non-hardening paint like fingernail polish or Glyptol.

Careful maintenance and the use of quality tape will prevent premature wearing out of the head. It is impossible to predict the lifespan of a playback head because of differences in materials used in manufacturing (permalloy heads are reported to outlast ferrite heads), variables in the composition of the tape (finely ground ferrite particles in some newer tapes cause less wear than older compositions), and differences in tape speed and pad pressure.

Owners of inexpensive equipment will probably want to buy a new machine by the time the playback head is worn out; most replacements will be higher priced equipment. Obviously, machines which use separate playback and record heads will get at least twice the playing time of combination head equipment before head replacement is necessary.

Cleaning and servicing playback heads: some shortcuts

You can save time by running cleaning tapes through cassette, cartridge, or open-reel equipment to remove oxide particles which

Figure 8-4. How to clean heads with a swab.

build up on the heads. For a more thorough job, use spray cleaner or denatured alcohol and cotton swabs.

Demagnetizing can also be done with packaged combination cleaner/demagnetizer tapes. Use them on every tape recorder/player that comes into your shop. Both you and your customers will often be amazed at the difference in sound obtained by simply cleaning the playback head—and you can do it without taking the machine apart!

Figure 8-5. Clean and demagnetize heads with a packaged tape. (Courtesy of Ampex)

You can also make a striking change in the sound of many players (especially portables which are subject to much rough handling) by spending a few minutes in aligning the heads (height and azimuth). Your customers will be pleased!

Fixing playback components

Use a signal tracer on a signal injector to track down the

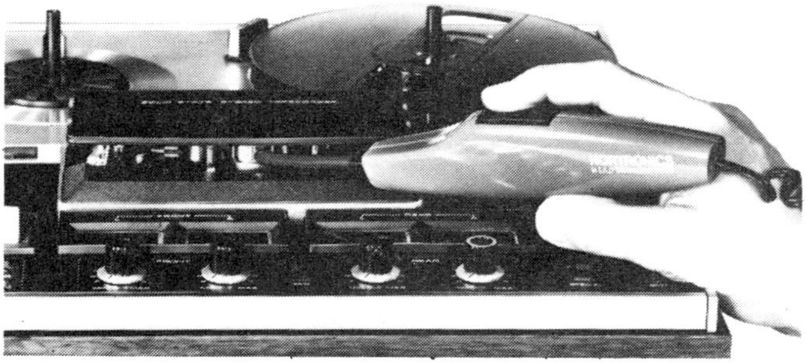

Figure 8-6. A handy head demagnetizer. (Courtesy of Nortronics)

trouble in an amplifier just as you would do in servicing other equipment. Tape player amplifiers are usually electronically simple but often have inaccessible parts.

Once you pin down the defective stage, check the components. Look for shorted or open transistors; use standard procedures to see if resistors are maintaining their values; check capacitors to see if they are deformed. Replace defective components.

In addition to repairing or replacing defective parts, you often need to adjust them. For example, you will need to adjust equalization on some machines. Demagnetize and clean the heads. Use a standard alignment tape which contains two frequencies—one recorded 10 dB below the other. Set the output from the lower frequency tone to read 0 dB on the VTVM, and then play the higher frequency tone. Move the playback equalization adjustment until the second tone is 10 dB below the operating level.

Although you may on rare occasions have to adjust the peaking coil or the play level, you will need to refer to manufacturers' manuals for correct procedures and specifications. These vary greatly. Generally, your repair time will be spent doing more routine tasks—checking for cold solder joints, grounded leads, or malfunctioning components.

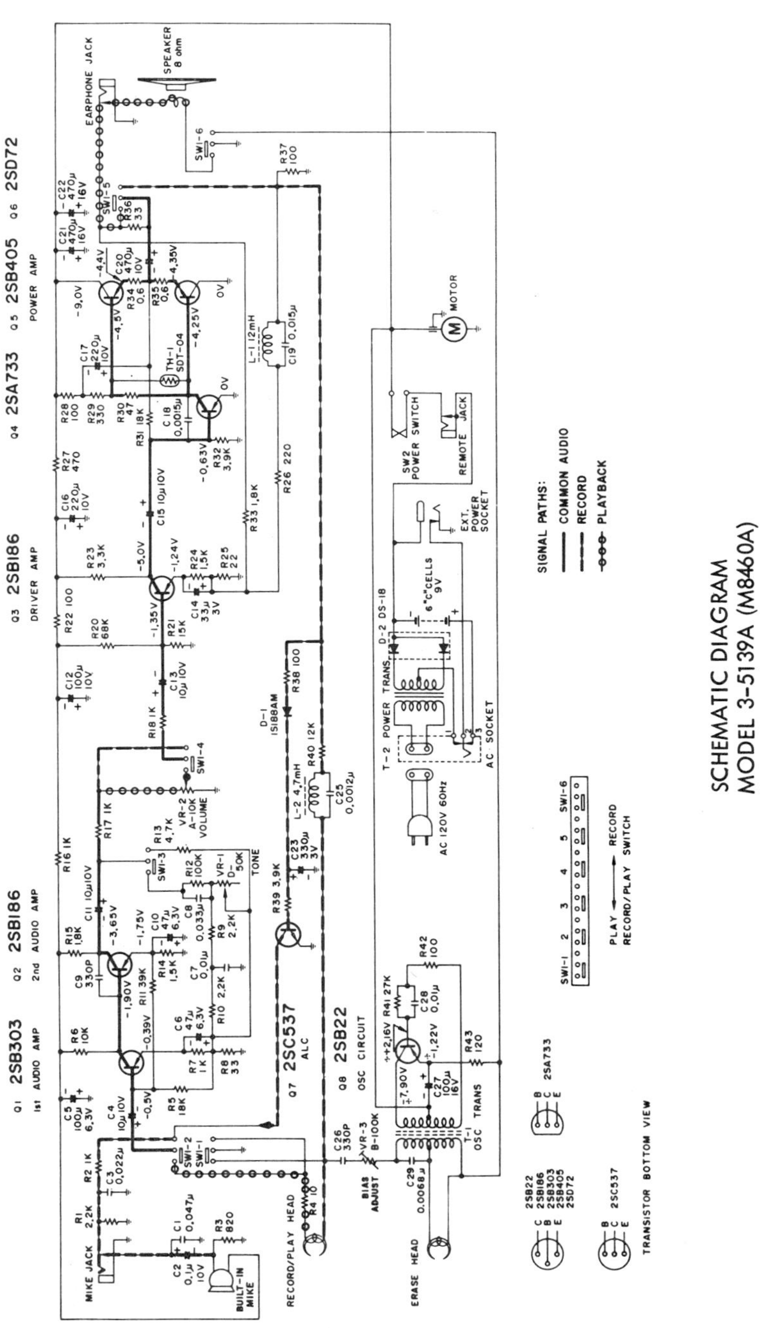

Figure 8-7. Schematic diagram. (Courtesy of General Electric)

Chapter 9

Maintaining Erasing Equipment

A major advantage of tape recording is that tapes can be erased and reused.

The way that tapes are erased

Magnetic signals on tape are erased by putting them into a magnetic field which rearranges the magnetic particles into random order so that they neutralize each other.

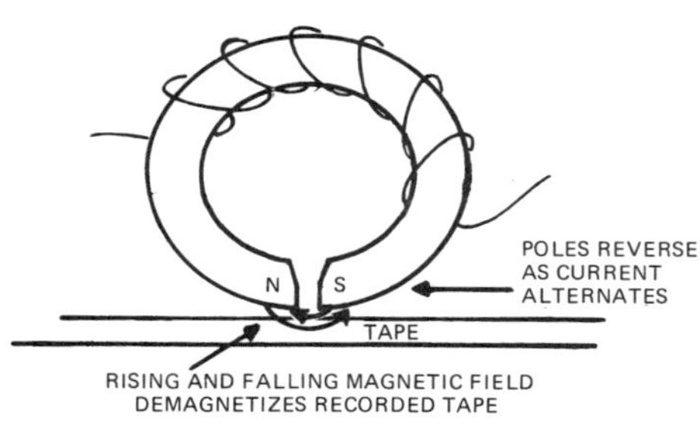

Figure 9-1. An erase head in operation.

This is done by an erase head or by a bulk eraser. The erase head is an electromagnet with a gap wider than that of the playback or record heads (sometimes fifteen times as wide), which is driven by the bias oscillator using alternating current. Turning the tape machine into record mode automatically engages the erase head so that the tape is erased before it reaches the record head.

The bulk eraser creates a broader magnetic field which per-

Figure 9-2. How to use one type of bulk eraser. (Courtesy of Nortronics)

mits the erasure of recorded material on an entire reel, cassette or cartridge all at once. Instead of running the tape through the machine past the heads, which could take 120 minutes or more (each side must be erased separately), the bulk eraser completes its job in a matter of seconds.

Better results are achieved by bulk erasers—partly because there are no alignment problems as in machine erasing (a tape that is a few millimeters higher or lower than the core of the erase head is not fully erased) and partly because the tape is erased a second time after leaving the bulk eraser when it passes through the machine in record mode.

Sections of a tape must be erased with the erase head in the machine, of course, although small errors can be removed with a tape editing pen (a small electromagnet with a quarter-inch active tip area).

You can thus erase a tape completely with a bulk eraser or partially with the in-machine head eraser, and make minute corrections with the tape editing pen.

Checking heads for wear

Since the erase head is manufactured with a wide gap, additional wear does not create the problems it does with playback and record heads. You'll seldom, if ever, need to replace worn-out

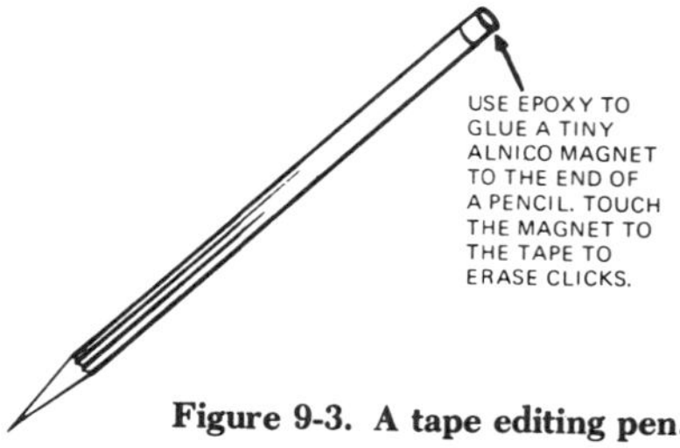

Figure 9-3. A tape editing pen.

erase heads. Damaged, shorted or open heads are not uncommon and, of course, they can be replaced.

Use a magnifying glass to see if the head is scratched or rough surfaced. Another clue is magnetic particles piled up like sawdust under the head; this is rare, but could be a cause of damage to valuable or irreplaceable tapes.

If your customer complains that tapes are not erased, the cause is usually other defects which will not be corrected by replacing the erase head. However, erase heads often are in need of alignment.

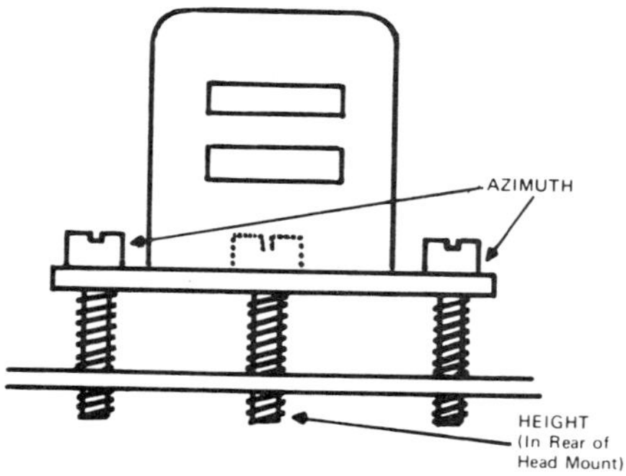

Figure 9-4. Erase heads must be aligned the same as record/play heads. (Usually adjustment is made with screws.)

Replacing and aligning heads

If you find it necessary to replace a shorted or open erase head, the process is simple, although you may find it difficult to get an exact replacement (check wtih your parts distributor). Unsolder or unplug the old head (after making a sketch of the wiring) and install the new one.

Alignment is also simple. Insert a tape (you may have to prepare a special cartridge or cassette by cutting away a portion of it—see illustration). Adjust the screw on the head or the guide so that the tape just covers the head core and extends no more than one tape thickness above it. You must also adjust the tape so that it moves parallel to the erase head and not at an angle to it.

Servicing dirty heads

Clean the erase head as you would the playback or record heads—on some machines the same head performs all three functions.

Use open-reel, cartridge, or cassette head cleaners which you simply run through the machine, or use a spray or bottled cleaner. Avoid scratching the head or damaging the leads connected to it. Some plastics are easily harmed by head cleaning fluids, so do not saturate the erase head. A rule of thumb to suggest to your customer is to clean the heads after ten to fifteen hours of play.

What to do if it won't erase

Customers sometimes misunderstand how to use their equipment. They frequently forget that missing interlock tabs on cassettes, for example, will not permit the erase/record mode to operate (cover the tab holes with transparent tape to correct this). Check to see that erase problems are not the result of putting the equipment in the wrong mode.

1. *If it only partially erases*
   a. clean the erase head.
   b. correct the alignment of the erase head.
   c. repair shorted or broken leads.
   d. replace the open or shorted head.

2. *If it doesn't erase*
   a. correct the bent erase head bracket.
   b. clean the dirty erase head.
   c. repair the shorted or broken leads.
   d. check the bias oscillator circuit.
   e. replace the defective record mode switch.
   f. replace the open or shorted erase head.

Check the erase head to see if it is shorted or open by using your ohmmeter.

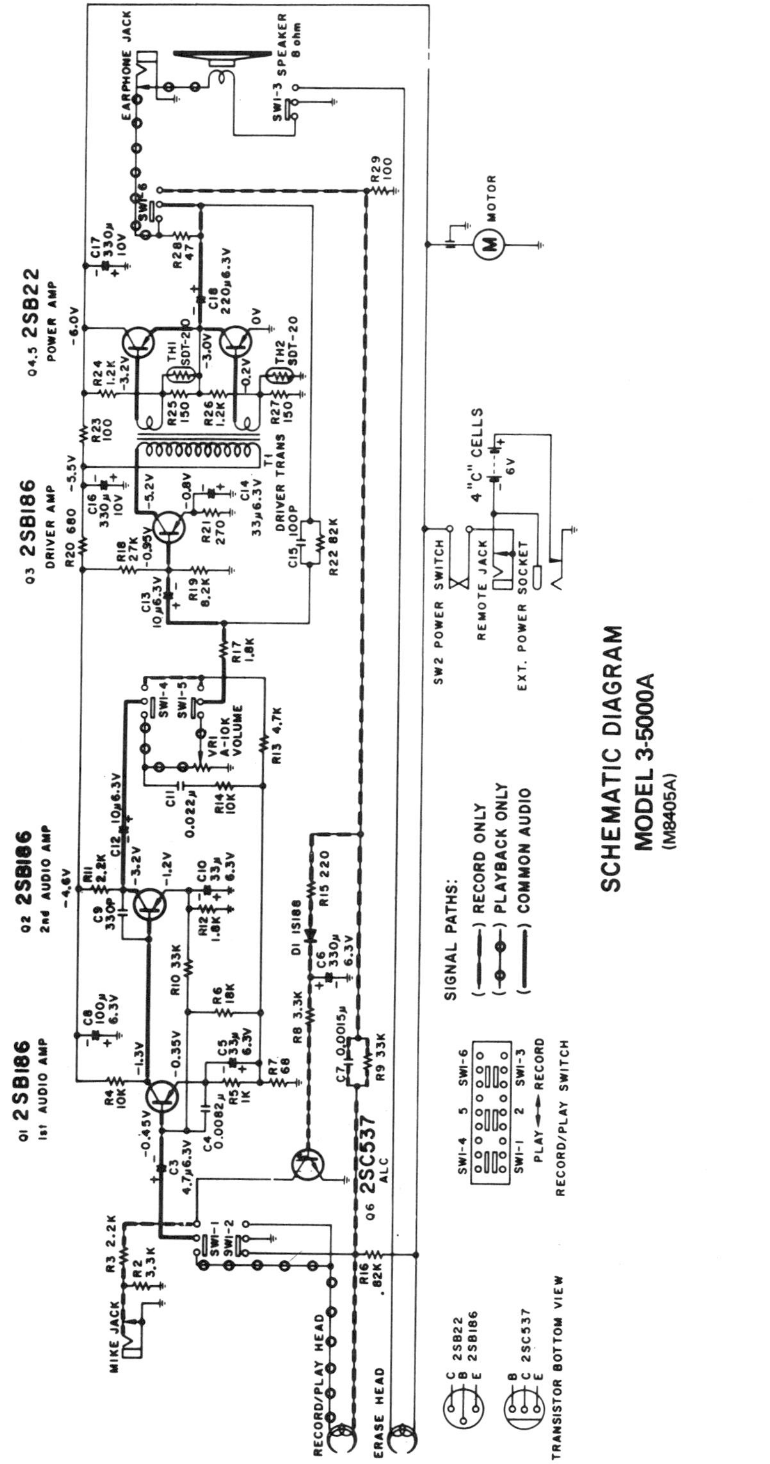

Figure 9-5. It is easy to trace the erase head connections in this schematic. (Courtesy of General Electric)

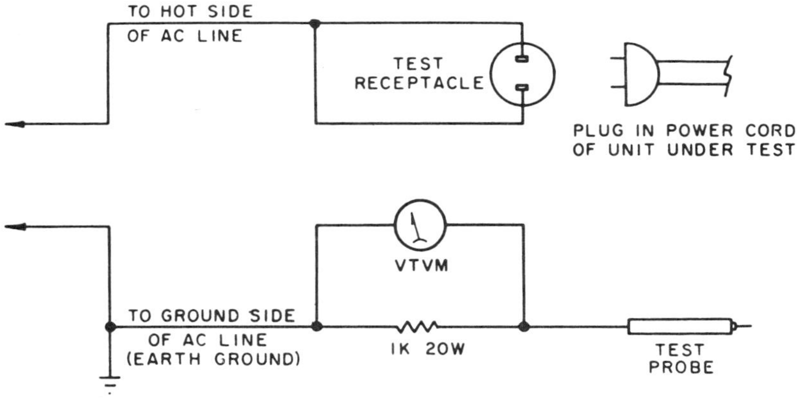

1. Connections to AC line must be as shown. Test probe must be at earth ground potential for OPERATOR safety.

2. Remove all externally connected test equipment and wires before safety testing this unit.

Construct circuit as shown:

3. Plug power cord of unit into TEST RECEPTACLE.

4. Switch unit under test to ON position.

5. Connect VTVM across 1K resistor in test circuit. Set meter on high (150 VAC) scale to avoid meter damage and touch the following points with TEST PROBE:

   a. Screw in A.C. Cord Compartment
   b. FUNCTION Knob
   c. EAR Jack
   d. MIC Jack
   e. Strain Relief Bracket

   If reading indicates less than 3 volts on all test points, proceed to a low (3 VAC) scale and repeat test.

6. Any reading greater than two tenths (0.2) volts indicates a potential shock hazard. If this occurs, further troubleshooting must be performed to determine the source of leakage.

Figure 9-6. For safety's sake, perform these tests on every tape machine! (Courtesy of General Electric)

A most annoying problem will be to record a program and to hear another program faintly in the background on playback. This can occur when the erase head is not correctly aligned. However, it can also be caused by recording a previous tape at too high a level (and then not erasing it fully) or by a phenomenon known as *print-through.* Since tape is very thin, loud sections may transfer their magnetism through the tape to sections which are soft (like a flute solo, for example). Nothing much can be done about the problem except to record carefully so that the tape is not overloaded at any point. (Tell your customers to watch the VU meter during recording and to ride gain). Careful storage (as described in Chapter 10) will also limit print-through from adjacent tapes. Customers often mistake this phenomenon as an erasing defect and you will have to explain it to them.

Erasing tapes out of the machine with bulk erasers

Bulk erasers permit you to erase a tape in a few seconds. Some require you to move the eraser slowly across each side of the reel; some permit the reel or cartridge to be placed in or on it and rotated; some perform the entire process automatically.

You can add to your profits by letting your customers erase their tapes free (or for a modest fee—usually ten tape erasures for a dollar) as a means of bringing them into your shop. The cost is nil, yet customers will value your service.

Chapter 10

Servicing, Editing and Repairing Tapes

Three types of tape carriers are common: the open reel, the reel-to-reel package type cassette, and the three-track cartridge (an endless loop on a single reel). Cassettes and cartridges can be inserted into the tape machine without threading. Cartridges, unlike cassettes (or open reel tapes) cannot be rewound or operated in the fast forward mode.

How magnetic tape is manufactured

Magnetic tape manufacture begins with the milling of the magnetic gamma-ferric oxide (Fe_2O_3) particles and their dispersion into a binder (usually vinylite). The material is coated onto the magnetic tape base (acetate, mylar, or PVC) and the magnetic particles are aligned (technically called being "orientated"). The material then is dried in an oven and slit into the required width. The finished tape is then put on reels or into cassettes or cartridges.

You need to know this because poor manufacturing techniques can produce bad tapes which can cause service problems. For example, if the oxide particles are not ground fine enough, they wear down the recording and playback heads abnormally fast; if they are not evenly dispersed into the binder they will not coat the tape evenly, causing dropouts or uncoated spots. If the tapes are not slit evenly, varying widths will cause the tape to skew as it goes past the heads and uneven recording and playing results will occur. If the slitting process is carelessly done, the resultant debris will

Figure 10-1. Tape is coated in wide sheets. (Courtesy of Capital Magnetic Products)

collect on the tapes and be deposited on the recording to playback heads.

Further, poor mixtures or misapplications of lubricants will cause poor contact with the driving capstan and the tape will go through the machine unevenly.

Understanding different types of tape

Most tapes produced today use one of three backings: acetate, mylar plastic or polyvinyl chloride (PVC). Mylar is the most popular because it resists breaking (it is two to three times stronger than acetate), ten to twelve times more dimensionally stable and eight to ten times as resistant to cupping, curling and absorption of humidity.

Figure 10-2. Tape is slit into appropriate widths. (Courtesy of Capitol Magnetic Products)

Selecting open reel tapes

Acetate tape (the cheapest) is adequate for most home uses but it tends to dry out rapidly and become brittle. Although this is not a problem in short term use, if you or your customers want to store memories of the baby's first word or a description of a wedding, use Mylar tape (which is more expensive). However, since Mylar stretches, the transport mechanism on inexpensive equipment may damage it. (Breakage is preferable, however, because it is more easily repaired.) Also, the thinner base may create print-through (the signal spreading from one layer through to another). Print-

Figure 10-3. Magnification of the coating on a top quality tape. (Courtesy of 3M)

through is most objectionable in voice recordings, although it occurs most frequently in loud passages of music—usually you won't hear it because of the phenomenon of persistence of hearing.

Mylar is much thinner than acetate tape; therefore, more can be placed on the reel. For example, a seven-inch reel contains 1200 feet of 1.5 mil acetate tape, which will permit you to record for 30 minutes at 7-1/2 ips (or one hour at 3-3/4 ips). Furthermore, you can obtain a reel of 3600 feet, which at slower speeds (1-7/8 ips, for example) will permit you to record for twelve hours or at 15/16 ips for 24 hours!

Reel-to-Reel
Recorded Both Directions

	15/16 ips	*1-7/8 ips*	*3-3/4 ips*	*7-1/2 ips*
150 ft.	1 hour	30 min.	15 min.	7-1/2 min.
300 ft.	2 hours	1 hour	30 min.	15 min.

	15/16 ips	*1-7/8 ips*	*3-3/4 ips*	*7-1/2 ips*
600 ft.	4 hours	2 hours	1 hour	30 min.
900 ft.	6 hours	3 hours	1-1/2 hours	45 min.
1200 ft.	8 hours	4 hours	2 hours	1 hour
1800 ft.	12 hours	6 hours	3 hours	1-1/2 hours
2400 ft.	16 hours	8 hours	4 hours	2 hours
3600 ft.	24 hours	12 hours	6 hours	3 hours

The selection of tape depends on the sophistication of your equipment as well. It is usually pointless to purchase a 3600 feet reel and then use it on a machine that has only speeds of 3-3/4 ips or 7-1/2 ips, for at those speeds you would get either three or six hours of play.

A cursory examination will reveal much about the quality of a reel of tape. What may appear at first to be equipment failure may be the result of poor quality tape. Examine the reel to see if it is warped. Insert a pencil in the reel and pull the tape. Does it bind or catch? Does it wobble off the reel? If so, the problem may be a cheap unbalanced reel or an inadequate method of splitting the tap which causes the tape and sound to be distorted.

In general, use and recommend name brand tapes. Obviously, most reputable manufacturers maintain quality controls; manufacturers of off-brand products often do not.

Cheap tape

Don't disregard cheap tape altogether, however. For some applications, it is fine. There is no reason to buy fine quality tape for a child's recorder or for a one or two time use. In general, using cheap tape for specialized purposes and for limited periods of time may not be bad—continued and prolonged use, however, may damage the tape machine's heads.

Selection of cassettes

Since all cassette recorders use only one speed, a 30-minute cassette tape (C-30) will record for slightly more than 15 minutes each side (tape manufacturers allow somewhat more tape so that

Figure 10-4. Spotless manufacturing facilities and great care go into manufacturing tape. (Courtesy of Ampex)

you won't be frustrated by having a few seconds cut out of whatever it is that you are recording). Tapes are commonly available in 30, 45, 60 (most popular length), 90, 120 and 180 minute sizes. It is also possible to buy tapes in length increments of two minutes: 30 minutes, 32 minutes, 34 minutes.

A 180 minute long tape seems like a bargain. However, many users of tapes of this length complain about jamming. Although these longer cartridges are ideal for recording longer musical works (symphonies or operas), use them on equipment which has a strong transport mechanism; the motors of most portable recorders are not strong enough to drive the tape past the recording and playback head in a steady manner. Your customer may be overloading his cassette player; your job will then be to explain this to him when he thinks his equipment is malfunctioning.

Although most cassettes look alike at first glance, there is a world of difference amongst them. One sign of quality is that the case should be fastened with screws so that a damaged cassette can be taken apart. Further signs of quality are rotating tape transports

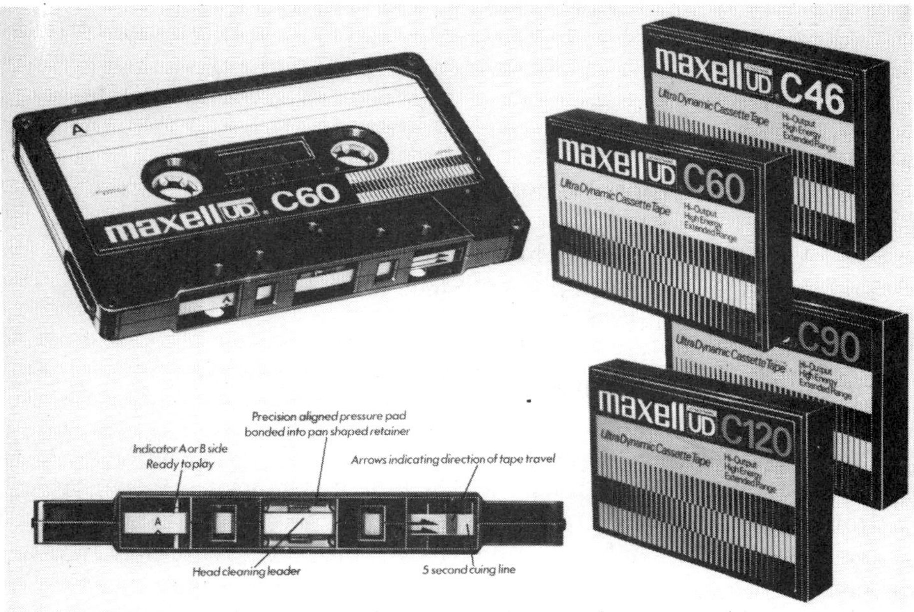

Figure 10-5. What to look for in a fine quality tape. (Courtesy of Maxell)

instead of cheap pins. Top quality cassettes use double flanged rollers (self-lubricating) to assure low friction. The plastic housing should be sturdy enough so that heat will not cause it to warp out of shape.

Further refinements are heavy-duty mu-meta shields which reduce stray magnetic fields and thus cut down on hiss levels, self-lubricating and non-binding hubs to assure smooth movement of the tape through the transport mechanism, and a well-balanced beryllium copper pressure pad to assure maximum pressure for head contact.

Cartridges

Most cartridges are commercially made, but equipment is available for recording and playback of eight-track cartridges. At present, cartridges are available only in 45-minute, 60-minute and 90-minute lengths. The cassette is increasing in popularity, but the cartridge still has an edge in that it can record/play eight tracks for better fidelity.

Specifications for cartridges are similar to cassettes. However, most cases are glued together or fastened under pressure at the factory rather than being held together by screws. Of all three types of tape carriers, cartridges are the most difficult to judge. The use of name brands guarantees some degree of quality.

Other types of tape carriers

A few manufacturers have produced cartridges and cassettes usable only in their equipment. Since production of such tapes is limited, there is little choice of types.

Fidelity in tape selection

Choosing tape depends on expected results. For example, it would be foolish to use a tape with a wide frequency response, say from 20 to 20,000 cps, if you were going to record a telephone conversation which has a much narrower range. On the other hand, if you were to record a symphony you would need as wide a range as possible. This too is tied in with tape speed. Use the formula 1.5 × tape speed = upper frequency response. For example, 1.5 × 7.5 = 1125 cps.

One type of tape may work better than another on a recorder if it is what the manufacturer's engineers used in originally setting the bias of the recording and playback head. Even though these may not be great, they can make some differences when there is a wide disparity between the tape and the recorder. Some recent machines have a dial to change bias. On others, it is technically possible to make different bias settings and to equalize them, but it is not worth the effort; the simpler solution is to use a more suitable tape, as you might switch to a different brand of gasoline.

How to spot defective tapes

Cupping, curling, dropouts and blocking are found in defective tapes.

Cupping means that the tape looks like a trough or a rain gutter. It is caused by materials which are not compatible to each other, much as a second coat of paint applied over another partially wet coat will not adhere correctly because of different drying rates. Slide the tape over the edge of a desk (coated side up) until it falls—cupped taped will overhang five to six inches or more before dropping; good tape will collapse almost at once.

Curling is caused by splitting the tape by a process which does not do it evenly. One side has thus had more stress than the other;

this distortion can show up in recording and playback. More importantly, it is a signal that the tape has not been subject to adequate quality control.

Dropouts are caused by improper adhesion of oxide particles to the tape so that spots which cannot be recorded appear. This also permits easy print-through. It is a further indication of poorly made tape. If possible, fold the tape and rub the edges together to see if it disintegrates.

Blocking means the tape will not separate from the reel when the reel is turned, causing the tape to unreel in jerks.

Layers sticking or adhering to each other (common after storing tape for a period), splices (tapes should be splice free), and short measure (since there is no simple way to measure a tape, an unscrupulous manufacturer can easily short you a hundred feet, thereby increasing his profits) are other tape problems.

Tapes which have an uneven coat will be noisy. Print-through is common and you will have to record at a higher volume, creating a vicious circle—the higher the volume, the more print-through; the more print-through, the more volume necessary.

Making repairs: splicing

The most common repair you will need to make on tapes is to splice broken ends. The process is simple.

Overlap the two ends and align the edges, making certain the glossy side of both ends faces up. Next, cut the overlapped ends at an angle. Throw away the short ends you have cut off; butt the tape ends. Splice the ends with splicing tape (cellophane tape and adhesive tape will bleed and ruin the rest of the tape). Trim the splicing tape beveling 1/64 inch into the edge of the recording tape.

How to replace damaged carriers

Although damaged open reels can be simply replaced (often a good cure for squeaking and squealing), cassettes and cartridges are slightly more difficult.

With a damaged cassette, separate the two halves by removing the screws or carefully splitting the case with a knife (inserting the knife more than 1/8 inch may damage the tape). Use parts from any easily available cassette repair kit.

In making a splice without a jig be sure to use de-magnetized scissors and follow this procedure:

1. Hold overlapped ends of magnetic tape securely between fingers and cut at shallow angle—about 45 degrees—for maximum strength and flexibility of splice.

2. Butt the cut ends of the tape in a splicing block without overlapping them. Make sure you apply the splicing tape to the shiny side of the magnetic tape.

3. Apply small section of splicing tape. Use the 7/32″ size and apply it parallel with the magnetic tape as shown. Rub splicing tape firmly with fingernail to iron out air pockets for positive adhesion.

4. If a splicing block is not available, lay magnetic tape on a flat surface. Butt the cut edges carefully and hold in position with index and forefinger. Apply short strip of ½″ or ¾″ wide splicing tape diagonally across splice and rub firmly to remove all air pockets.

5. If splicing tape is wider than the magnetic tape used, trim off excess, cutting into magnetic tape very slightly. This eliminates danger of exposed adhesive gumming up recording head or sticking to adjacent layer of magnetic tape.

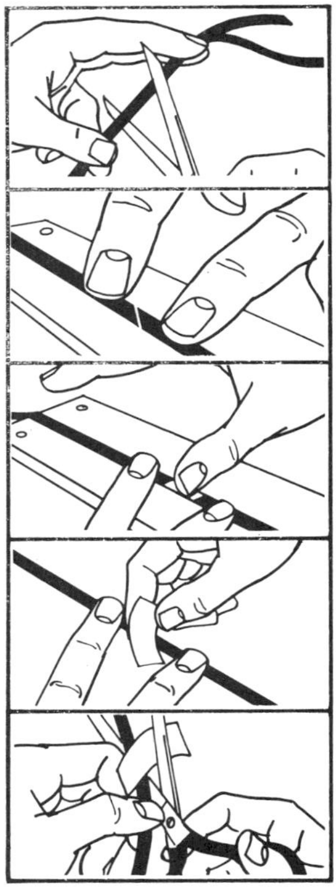

Figure 10-6. How to splice a tape. (Courtesy of 3M)

If the cassette has damaged hubs, cut the leader in half and splice the tape to the leader ends. Otherwise, shift the old hubs to the new cassette.

Install the old tape in the new cassette, reposition the pressure pad, position the top half of the case and fasten the halves together with screws.

Repair kits are also available for cartridges. Split the case carefully with a knife (avoiding inserting the knife more than 1/8 inch to avoid tape damage). Do not cut the tape. Instead, interchange hubs, threading the tape through the new carrier. Fasten the halves together with screws.

Keeping tapes in serviceable condition

Some authorities recommend that tapes not be rewound after play until the next time you wish to use them. The theory behind

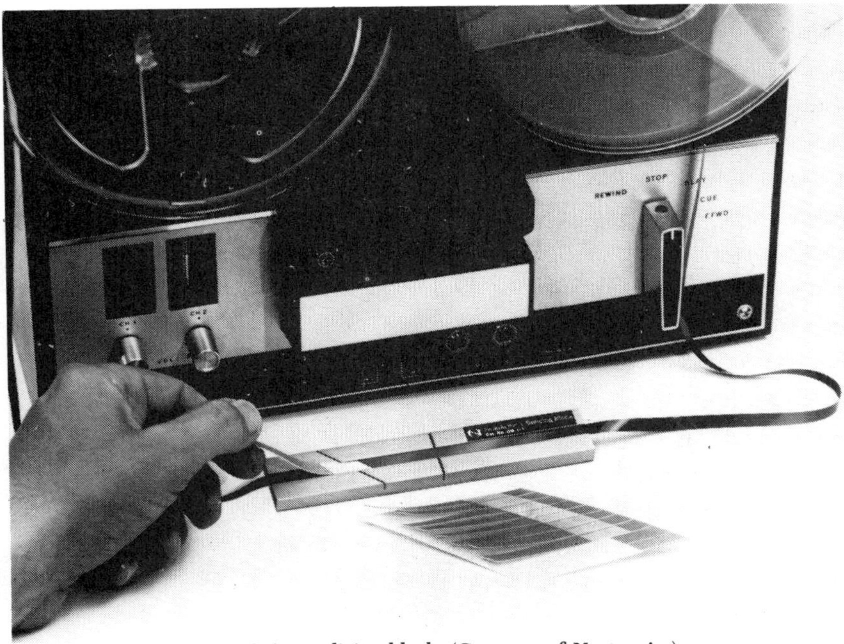

Figure 10-7. A splicing block. (Courtesy of Nortronics)

this is that play-through will be less likely—the tape will not stick to itself since it has just been rewound. If you follow this policy, do it with all your tapes so that you do not make an error and get confused in rewinding.

What at first may seem to be a problem with a recording may be the fault of the cassette or cartridge. A wow or jamming may be caused by an oversized cassette on a relatively weak motor or a cartridge that has warped after exposure to heat.

Some problems can be solved simply. A jammed cassette (caused by the mass of tape being too tightly compacted for the motor to turn) can often be freed by slamming it several times on a flat surface to loosen it. Insert a pencil into a cassette and wind or unwind it. In extreme cases, remove the screws and open the case.

Warped reels, cassettes and cartridges must be replaced.

Editing tapes

As a service technician, you may never need to edit a tape; you need to understand the theory, however. Editing means changing the tape by rearranging sections or combining them with other tapes. This can be done mechanically or electronically.

Mechanical editing entails cutting pieces of tape and splicing

Figure 10-8. Damaged cartridges can be easily repaired. (Courtesy of Ampex)

them in another place. It is useful in changing one track open-reel tape, but cannot be done with tape cassettes or cartridges. It leaves little room for error.

Electronic editing means re-recording the signal from one tape onto another in the desired order. This can be done with all types of tape and tape carriers. The process is simple. Connect one recorder to another. Play one and record on the other. A less effective method is to place a mike in front of the speaker.

Editing tapes requires planning. Use a stop watch and a chart to write down the correct order of the tapes.

Chapter 11

Using, Repairing and Maintaining Accessories

Increase the versatility of any tape recorder/player by using the wealth of accessories which are on the market. You will, in addition, increase your income if you can show your customers ways and means to use them more effectively.

How to select and use expensive accessories

Those accessories which cost lots of money are either complicated pieces of equipment or are very specialized. Your customers who use such equipment expect their electronic gear to be of top quality and their service technician to be knowledgeable in its use.

For example, *mixers* (which permit combining and balancing of the output of two or more mikes recording musicians and vocalists up to 500 feet away) are necessary for professional recording. A top quality machine may have four channels with individual faders for each channel, and such useful devices as separate equalization for high, medium, and low frequencies as well as group faders for each output channel. Fine equipment will have specifications of 40 Hz to 15,000 Hz plus or minus 1-1/2 dB with distortion no higher than 0.5 percent.

Other types of complex equipment have also become popular. If your customer doesn't have Dolbyized components, sell him a *Dolby noise reduction unit* (to cut tape hiss) which can be added to his present recorder/player. Most units are compatible with any recorder and allow simultaneous processing and deprocessing of signals. Good equipment will reduce tape hiss by 10 percent and

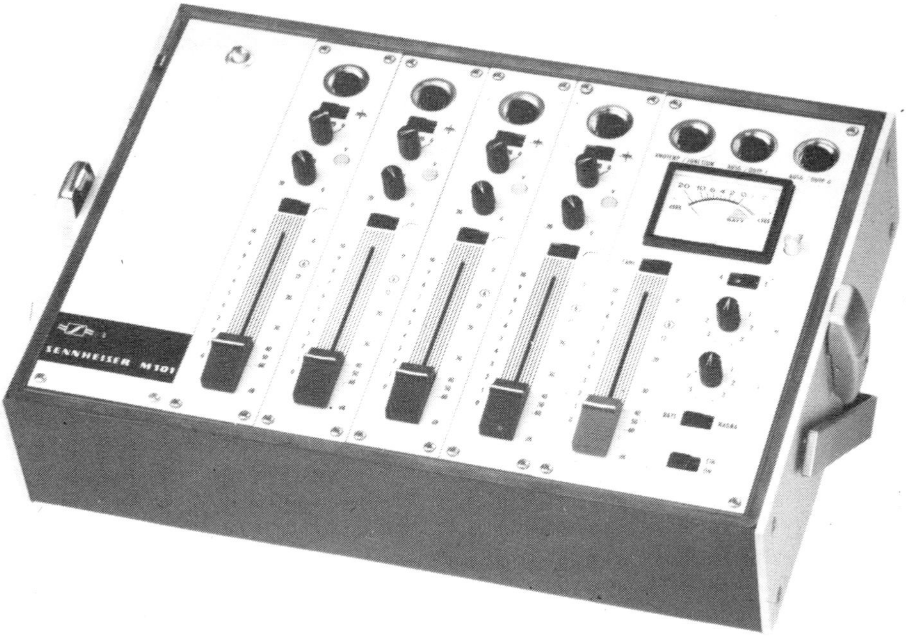

Figure 11-1. A mixer. (Courtesy of Sennheiser Electronic)

will have a frequency response from 20 Hz to 20,000 Hz plus or minus 0.5 dB.

Another useful unit is a *quadraphonic decoder or demodulator*. Most present quadraphonic recordings are discs, but Q-8 cartridges are also available. To convert a stereo system to quad sound (also called 4 channel sound), connect a commercially made SQ decoder or a CD-4 (compatible discrete) demodulator. Add two more speakers the same size (or smaller versions of the same model for the rear speakers) for a total of four speakers—one placed in each corner of the room. Compare the specifications of the quad equipment to see that it matches the original components in quality and performance.

In addition, you will find other accessories ranging from sound echo adapters to single channel equalizers—all of which are useful for specific applications. Such equipment will account for a very tiny portion of your service calls, however, even though you should be familiar with their selection and use. In almost all cases, it will be more profitable to send them out to the manufacturer or to a factory-authorized repair center than to service them yourself.

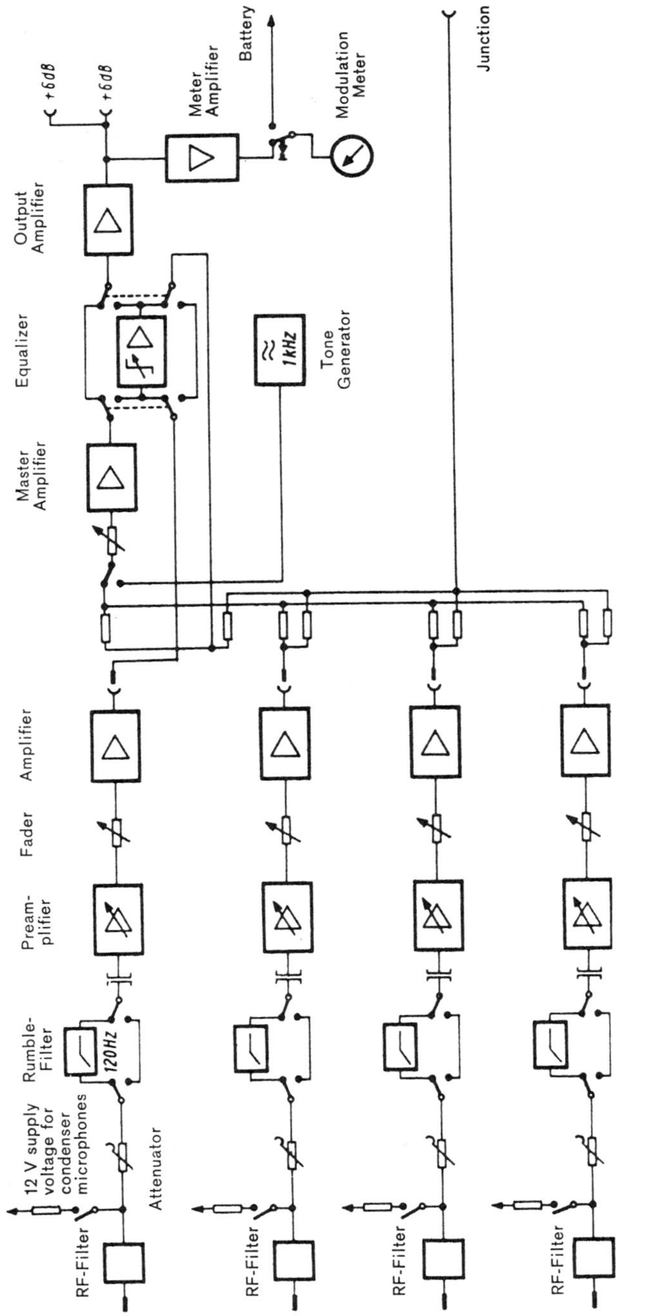

Figure 11-2. A block diagram of a fine quality portable mixer. (Courtesy of Sennheiser Electronic)

Choosing small accessories—mikes, headphones, cassette winders and others

A wide range of accessories is available to expand the use of any recorder/player. For example, devices are available which start and stop a recorder with a foot pedal; these are useful for typing dictation. A recorder can be connected to a telephone or to another recorder by cables which are readily available. Other units permit recording directly off the air (AM or FM) or from a record player.

Figure 11-3. A quality microphone. (Courtesy of Shure Brothers Incorporated)

The most common and useful accessories are microphones. Various mikes have specific advantages and disadvantages. Ceramic microphones are rugged and resistant to temperature and humidity changes but need higher amplifier gain than other mikes. Crystal microphones are highly sensitive, but are easily damaged by shock or extremes of temperature or humidity. Dynamic (moving-coil) mikes have low impedance which permits the use of long cables but they also have low output. Ribbon microphones are tough and dependable but are not good for use outdoors. Reluctance mikes are rugged and long lasting, but do not produce high quality sound.

So, what do you look for? Select a mike that best suits your purpose (see Chapter 12). Help your customer choose a mike with an extended response range for more faithful reproduction, with a response curve that is smooth (no abrupt peaks or valleys of output at any frequency) and flat (responds equally well at any frequency). Try to point out the values of using a dynamic microphone with a frequency response to 50 to 15,000 Hz which will fit the needs of most amateurs.

Microphone styles range from a miniature omnidirectional unit about the size of a button (worn by speakers who like to move about on the podium) to a large parabolic reflector used to make recordings of sound some distance from their source (bird calls, aircraft and train sounds, for example). These mikes are variations on the basic ones. You and your customers will seldom have use for them.

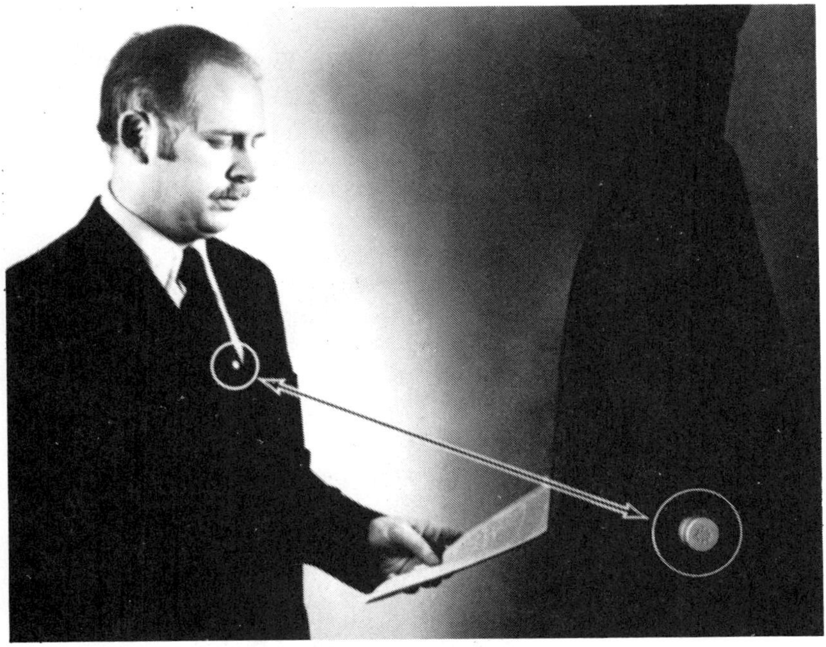

Figure 11-4. Button size microphone. (Courtesy of Electro-Voice)

Headphones are also useful accessories. Some headphones can be used to block out all other sound or to hide under a pillow for sleep learning; a lightweight phone can be inserted in the ear. Electrostatic headphones are even better than loudspeakers in

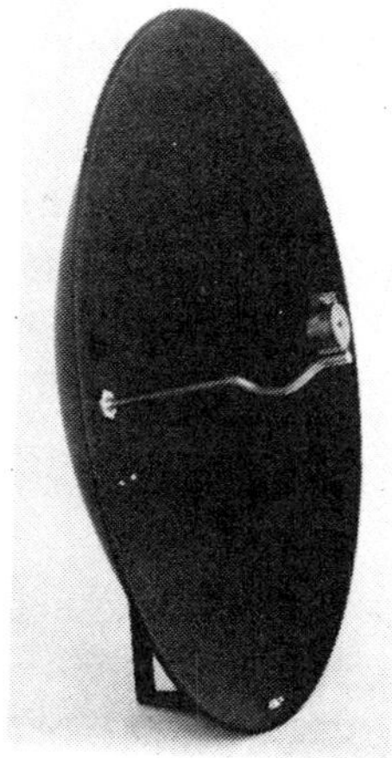

Figure 11-5. A parabolic reflector. (Courtesy of Ercona)

their smoothness and frequency range; they are, however, expensive and somewhat heavy.

In offering advice on selecting the best headphones, suggest that your customer use comfort as a first priority. Even if a set of phones comes with top rate specifications, if it is an ear pincher, it won't see much use. Assuming your customer finds a comfortable set, he should also expect approximately 16 to 20,000 Hz plus or minus 3 dB response.

One other inexpensive accessory that most recording fans find useful is a *fast-rewind device*. Use this to speed up rewinding tape

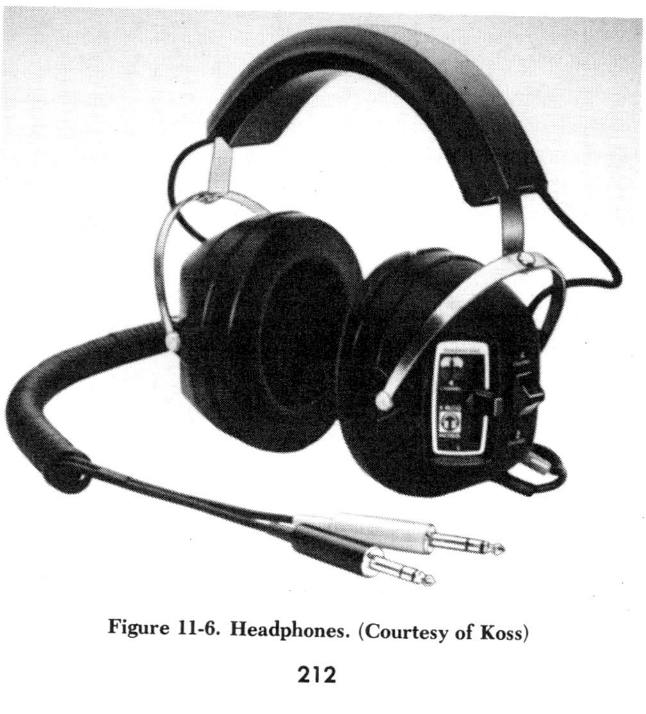

Figure 11-6. Headphones. (Courtesy of Koss)

out of the machine, to avoid tying up the use of the recorder/player, and to save wear and tear.

Repairs other than resoldering a loose joint or gluing a broken part are best left to the manufacturer or his authorized representative. You'll only waste your time otherwise.

A clear-cut guide to help you know when to repair and when to discard an accessory

Give expensive accessories a brief but thorough examination for broken leads and similar electronic damage which is visible to the eye. If nothing is obviously amiss, send the unit out to the manufacturer or to a local authorized repair shop. Such equipment is so specialized that it would be a waste of time for you to try to trace the trouble.

With inexpensive accessories, the rule of thumb is simpler still. Do not attempt any repairs other than soldering loose connections or gluing broken parts. Components for equipment of this nature are often custom made and unobtainable from your normal sources. If soldering or gluing won't fix the accessory, tell your customer that it isn't worth repairing and recommend that he purchase a new one. (You have a sure sale if you offer a slight discount—even as little as 10 percent—on that new purchase.)

MURADAPTER
Model CA-48

Figure 11-7. A stereo cassette adapter—one of many small but useful accessories. (Courtesy of Mura Corporation)

Chapter 12

Achieving Professional Recording Techniques

Add to your profits! Use your knowledge of tape recording to make professional recordings of singers, musicians and speakers.

Basic principles of top quality recordings

Professional recordings are crisp and clear (not muddy or fuzzy) and well balanced. Errors are edited out of the tape; corrections are added; fluffs are erased.

For simple recordings like those your customers will do at home, procedures are elementary. Advise them to speak about eight to twelve inches from the microphone (to avoid sound distortions like popping their "p" 's). Minimize background noises by recording in a quiet section of their home.

When recording outdoors, wrap the microphone in an old sock to minimize wind noise. Indoors, put the mike on a foam rubber pad to eliminate banging or thumping noises from handling.

Advise your customers to use top quality tape and to clean and demagnetize tape heads. Suggest that they record music at a faster speed than speech if their equipment permits such adjustments.

For professional recording, you need to understand ways to select and use microphones and methods of achieving correct balance, as well as the various techniques used to achieve unusual effects.

The correct placement of the microphone is one of the tricks of the trade that separates the pros from the amateurs. For example, if you are recording someone who lisps or hisses when he speaks, have him stand at an angle to the mike. This cuts down and often eliminates the hiss being recorded on tape.

Correct placement is also necessary to maintain correct balance. For instance, if a person is accompanying his own singing by playing the piano, you must place the mike about eight to twelve inches from *him*—otherwise the piano will drown out the sound of his voice. A small group of vocalists must stand three to ten feet from the mike with their backs to the piano. To make the group sound larger, simply move the mike farther away.

Better balance can be achieved by the use of two or more microphones, strategically placed, and combined by the use of a mixer. Thus, in recording a singer at the piano, two mikes would permit individual adjustment of the sound from the piano and the voice of the singer. The mixer permits you to highlight specific voices or instruments at will. It is a necessity when you want to record a soloist accompanied by a chorus.

Generally, you'll find that three mikes (one for the soloist and two for general balance) will improve the sound of any musical recording.

Careful selection of microphones can improve your recordings. *Unidirectional microphones* pick up sound from the front while suppressing sound from the back. This cardoid (heart-shaped) pattern suppresses rear sounds at least 80 percent, which makes it ideal for elimination of noise caused by an audience as well as feedback. A performer can work farther away from this type of mike than from other types.

Omnidirectional microphones pick up sound evenly from all directions. Use them for interviews and any application where feedback or audience noise is not a problem. *Bidirectional microphones* pick up sound from the front and back and suppress sound from the top, sides and bottom. Use these mikes for two performers on opposite sides of the mike, especially in rooms with hard ceilings (which create feedback problems).

You may find that you prefer the sound produced by one type of mike over that produced by other types. Eventually, you'll settle for a quality mike (probably either dynamic or ribbon type) with a flat smooth response over 50 to 15,000 Hz. The more extended the response, the more faithful the reproduction.

Sound-on-sound recordings and *sound-with-sound recordings* are studio techniques that you can use to create stunning effects, in

Cardioid

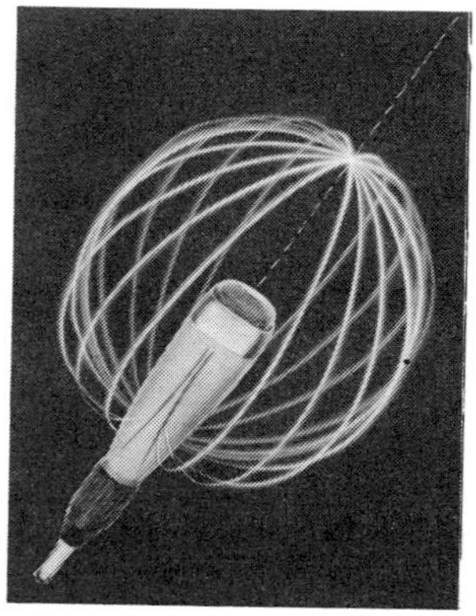

Omnidirectional

Figure eight

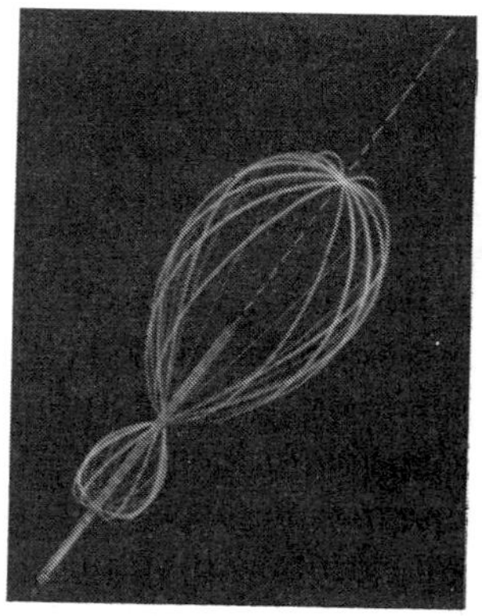

"Club-shaped"

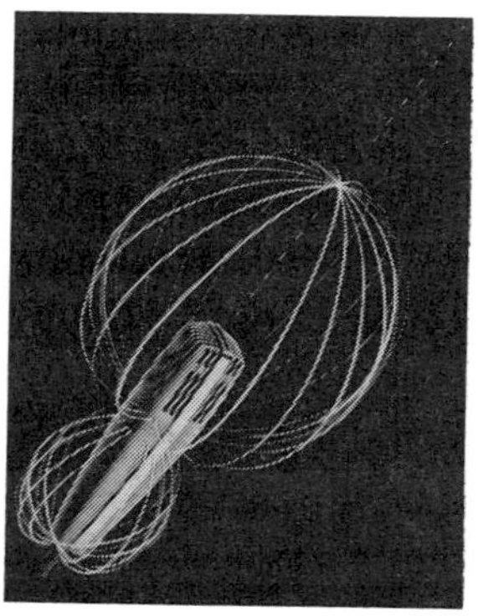

Super cardioid

Figure 12-1. Directions from which microphones pick up sound. (Courtesy of Sennheiser Electronic)

FREQUENCY RESPONSE

The fidelity of reproduction afforded by the microphone depends on three factors of the frequency response:

1. Response Range. In general, the more extended the frequency response of the microphone is, the more faithful the reproduction will be.
2. Smoothness. A high fidelity microphone is made so that no conspicuous abrupt peaks or valleys of output occur at any frequency. This results in an essentially smooth frequency response curve.
3. Flatness. A flat frequency response curve is one showing output remaining at approximately the same level throughout the frequency range. This means that the microphone responds equally well at any frequency, an essential of high fidelity reproduction.

While the microphone with the widest range, smoothest and flattest frequency response curve will give the highest fidelity, it is not necessarily the best microphone for every application. [For example, a shaped (peaked) response is often devised to achieve added "presence" for microphones used in paging and communications systems.] Frequency response required for various applications is summarized below:

RESPONSE REQUIREMENTS

Use	Response Range	Response Character	Possible Microphones
Recording, broadcasting, highest quality public address for music and voice.	50-15,000 Hz	Flat Smooth	Ribbon Dynamic
High quality public address for voice and music.	70-10,000 Hz	Flat or slightly rising No prominent peaks	Ribbon Dynamic Some Ceramics
Indoor paging systems and public-address systems for voice use.	200- 5,000 Hz	Slightly rising No prominent peaks	above plus Some Controlled Magnetics
Outdoor paging systems or systems in noisy locations. For voice use when maximum intelligibility required.	300- 3,500 Hz	Rising A peak in the 2000 to 3500 Hz region is often desirable	Controlled Magnetic Dynamic Some Ceramics

Figure 12-2. Frequency Response (Courtesy of Shure Brothers Inc.)

addition to those created by microphone selection and placement. Sound-on-sound recordings use two or more voices or other sounds superimposed on one track. This *monaural* recording is created by playing material on track 1 while at the same time recording both that material and new material on track 3. The process is then reversed so that the material on track 3 and new material are recorded on track 1. Thus, you can continue this process as much as you like, creating, duets, quartets, and choruses with only one person.

SOUND-ON-SOUND OPERATIONAL SEQUENCE

		1st Recording	2nd Recording	3rd Recording	4th Recording
Connections (Mic, FM Tuner, etc.)		LEFT	RIGHT	LEFT	RIGHT
RECORDING LEVEL Control Knob	LEFT	Adjust the input level of the 1st recording	Minimum position	Adjust the input level of the new recording	Minimum position
	RIGHT	Minimum position	Adjust the input level of the new recording	Minimum position	Adjust the input level of the new recording
Position of Sound-On-Sound Switch		OFF	L R	R L	L R
Position of MONITOR Switch	LEFT	TAPE	SOURCE	TAPE	SOURCE
	RIGHT	SOURCE	TAPE	SOURCE	TAPE
SOUND-ON-SOUND Level Adjustment Knob			Adjust the input level of the recording supplied from track 1 to track 3	Adjust the input level of the recording supplied from track 3 to track 1	Adjust the input level of the recording supplied from track 1 to track 3
RECORD Button Used		LEFT	RIGHT	LEFT	RIGHT
Track Used		Track 1	Track 3	Track 1	Track 3

Figure 12-3. Sound-on-sound operational sequence. (Courtesy of Bell & Howell)

Sound-with-sound is a stereophonic technique. This method, useful for language lessons, requires you to record on track 3 while playing track 1. Then these two tracks are played back at the same time in stereo, sounding as if they were recorded simultaneously.

Figure 12-4. Microphone. (Courtesy of Shure Brothers Incorporated)

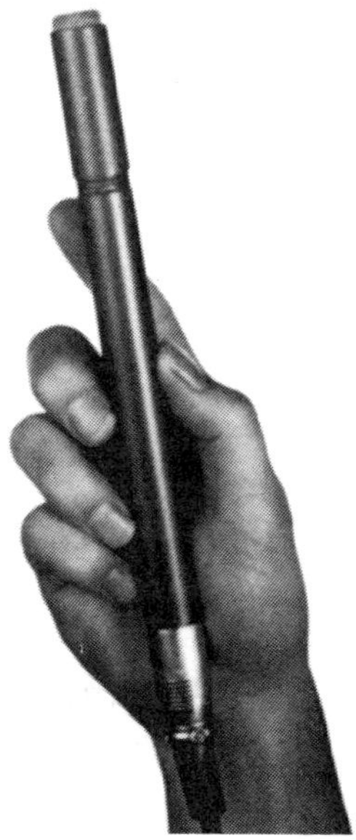

Figure 12-5. Hand-held microphones pick up more noise than those mounted. (Courtesy of Shure Brothers Incorporated)

In addition to using these techniques, you may wish to record from *records (discs), radio,* or *television*. You'll seldom encounter difficulty in recording 33-1/3 rpm discs, but if you try to preserve the sounds on those antique 78 rpm versions of Tommy Dorsey, you're bound for some headaches. Try spraying them with a cleaner to cut out some of the surface noise (clean really dirty discs with detergent under running tap water) and use an ordinary crayon to help smooth out scratches. Obviously, use a filter system if one is available.

When recording from *radio* or *TV*, you can make recordings by placing the mike next to the speaker, but you'll never be happy with the results (which will be muddy and fuzzy). Instead, if there is no jack for a direct connection, attach a shielded cable to the voice coil terminals of the speaker. Adjust the level to avoid overload.

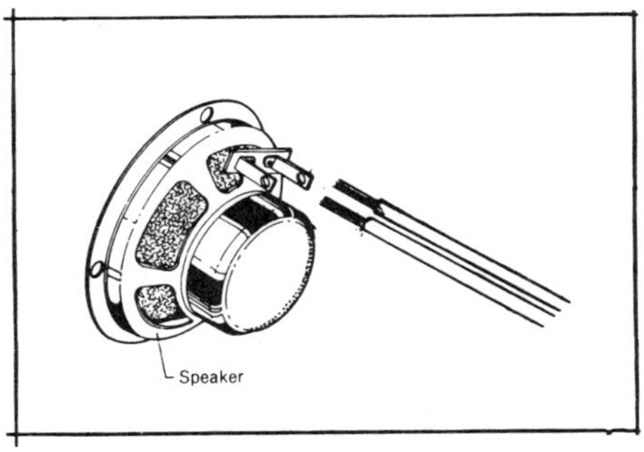

Figure 12-6. How to connect a cable to the speaker output terminals. (Courtesy of Clarion Corporation of America, Los Angeles, California)

Ways to measure audio quality

Without question, you can measure audio quality with your ear—ask yourself if the sound is muddy, distorted or tinny. However, what might seem like quality sound may be defective or distorted in subtle ways; you will only come to know this when you have listened a long time. You will actually get fatigued by listening to distorted sound!

Luckily, you can measure the quality of sound by using technical equipment. The easiest way to do this is to make quality checks on level and balance as you record. Watch the needle on the VU meter to see that it stays within bounds. Although some old timers can check balance by standing at the mike with a hand covering one ear, you'd be wise to use a meter. Check the signal coming from each channel on your mixer or each of your mikes. See that it stays within limits.

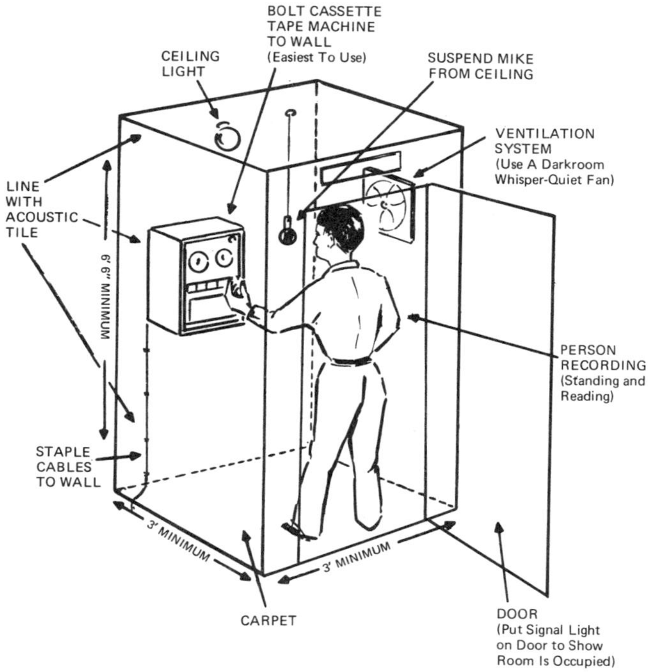

Figure 12-7. A studio in a closet.

Building a simple sound studio

Three basic types of sound studios can be built in your shop to boost your profits at minimal cost and time.

First, you can simply install recording equipment in a small closet or in any corner that is relatively quiet. Beware—your shop has its share of street noise, radios, TVs, and recorders booming away, as well as the racket caused by you and your customers! To cut out this unwanted noise, you will have to soundproof your studio with acoustical tile (which is stapled into firring strips) and you must put a thick carpet on the floor. Add a unidirectional microphone (with its back towards the door or the source of most noise) and run the cables in an out-of-the-way spot so no one trips over them. This arrangement will be adequate for limited use by one or two people.

A larger studio will create larger profits. Rock music groups are good customers for the facilities and skills you can offer them if you have an understanding of their needs. First, they usually need a lot of room for their equipment and second, extensive electrical connections. Convert an entire room into a recording studio. To do so, you need to build a room within a room. Mount two-by-fours as studs to form a new wall a foot inside the old wall. Leave no windows or other openings. The ceiling should also be lowered and ideally, the floor should be suspended (you can get adequate results from using a heavy rug in most cases, however). The walls should be covered with acoustical tile or other sound-absorbing materials. Curiously, you won't want to dampen out all sound —absolute silence transmits such tiny sounds as the beating of your heart!

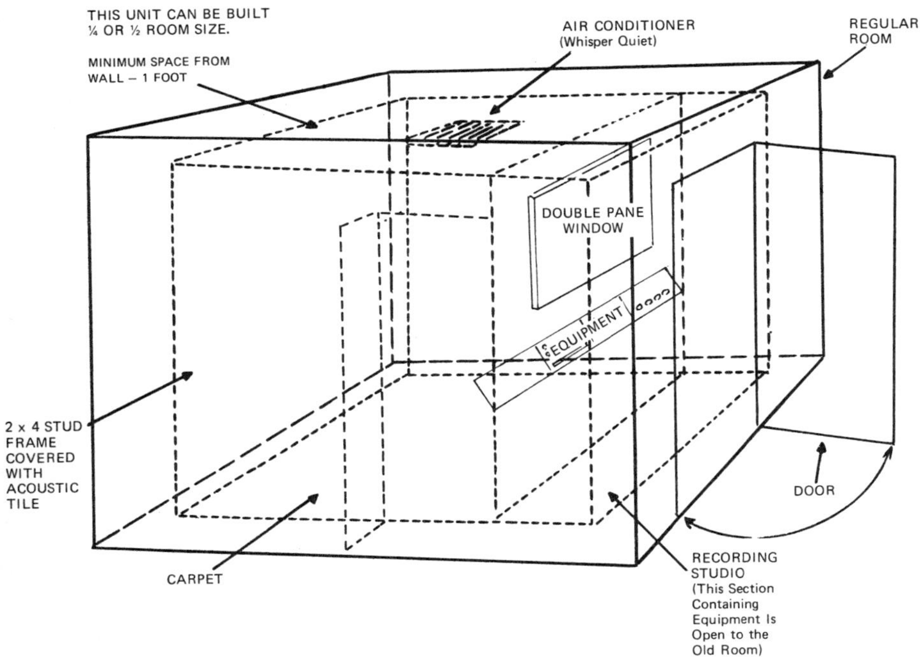

Figure 12-9. Professional studio operation. (Courtesy of Sennheiser Electronic)

Inside the room, you might build a closet to hold the recording equipment, with a window looking out on the studio. In this way the noise of the electronic equipment would be kept from the mikes.

Since building a good studio is such a complicated process, you may wish to buy a prepackaged studio, engineered to produce professional results. These prefabricated studios need only to be assembled to put you in the professional recording studio business!

Specify strict rules. Be sure that users understand that your charge is on an hourly basis ($20 an hour or more, if you must hire a technician), and that you charge an extra hour for any portion of an hour. You can also boost your profits by making duplicates of tapes recorded by groups who use your studio.

Copying machines (high speed and one-to-one) and how to use them

Copying machines are available in sizes ranging from desk-top units for cassettes to floor-sized models for open-reel tapes. Such equipment is fast and relatively simple to operate.

Cassette copiers make copies of a 60-minute tape (both sides at

Figure 12-10. A high-speed cassette copier. (Courtesy of Telex)

once) in less than two minutes! Some machines use add-on modules to make up to five copies simultaneously.

One-to-one machines make one copy of a tape at a much slower speed: the time it takes the tape to play—15 minutes, 30 minutes, 60 minutes. These machines are useful only if you are making small quantities and do not need to reproduce them quickly.

Ownership or rental of such equipment can open up new profits for you. In addition to copying tapes recorded in your studio, you will find that nearby schools, businesses and similar institutions will let you do their copying for them. Individuals will seek you out to get copies made. Copying charges usually average about 50 cents a tape copy.

Obviously, there are restrictions. Copyright laws prohibit you from copying commercial tapes on which royalties are paid.

A simple program for keeping up with the latest developments in tape technology

To maintain your wizardry with tape recorders and tape recording, you need to keep one step ahead of your customers. You need to know what's new! Keep informed—subscribe to at least one electronics magazine and read (if nothing else) the column devoted to trade news. Find time once a year to visit a trade show to see what the new models look like and to load up on free advertising literature and information from the salesmen. Read the materials distributed by the manufacturers with whom you do business. Finally, set a goal of reading at least one new book a year on your favorite subject—tape recording!

Glossary

acetate—Cellulose acetate plastic film used as a backing for inexpensive tapes.

acoustic—The specific characteristics of a room which affects sound produced in it.

alignment—The position of a tape head to the tape running past it.

amplify—To strengthen an electrical signal by increasing its amplitude.

attenuate—To weaken an electrical signal with a volume control.

automatic reverse—Device incorporated into some recorders to permit playback of other tracks in a reverse direction.

automatic shutoff—A special switch incorporated into some recorders which stops the machine when the tape breaks or ends.

azimuth—The angle of a tape head in relation to the direction of the tape moving past it.

backing—The base or material used as carrier for the magnetic oxide coating.

balance—The relationship between the volume of two (or more) speakers.

bias—ac signal added during recording to compensate for the nonlinearity of magnetic recorders.

bulk eraser—A device used to erase an entire reel or cassette out of the recording/playback machine.

capstan—The spindle which drives the tape through the machine by pressing it against a pinch-wheel, thus keeping the tape speed smooth and constant.

cartridge—A tape carrier using a single-spool, endless loop 8 track tape.

cassette—A tape carrier using two reels (one for supply and one for takeup) in a closed package which can be inserted or removed instantly without threading.

ceramic microphone—Uses a ceramic component to convert sound waves into electrical energy, but requires higher amplifier gain than crystal microphones.

channel—A single and complete path through an electronic system.

coercivity—The demagnetizing force required to reduce induction to zero.

condenser microphone—Uses varying electrical charges across a capacitor for transduction.

cps—Cycles per second or Hz.

cross talk—Signals from one channel leaking into another channel during playback.

crystal microphone—Uses salt crystals to convert sound waves into electrical energy.

decibel—The relative measure of sound volume. One dB is the smallest volume change detectable by the human ear.

deck—A tape machine with no amplifiers or speakers, but designed to be a part of a large system containing those components.

distortion—A difference between the original sound and the sound reproduced through electronic equipment.

Dolby—The trade name for a noise reduction system which increases signal-to-noise ratios by raising the volume of quiet passages during recording and lowers the ratio during playback.

dropout—A loss of signal resulting from imperfections in the tape.

dub—To copy a recording.

dynamic microphone—Uses a moving coil in a magnetic field to convert sound waves into electrical energy.

dynamic range—The difference between the loudest and the softest sound that equipment can reproduce without distortion, expressed in dB.

editing—To select and rearrange recorded material to achieve an effect different from that originally produced.

equalization—To change the frequency to meet certain requirements.

erasure—To remove the recorded sound from a tape by placing it in a strong magnetic field of alternating polarity while reducing the strength of that field.

fast forward—A mode which permits the tape to move in a forward direction at a high speed to reach desired selections.

feedback—A squealing noise generated by a microphone picking up the sound emitted from a speaker from its own amplifier system.

fidelity—The measure of how close the sound reproduced by a system is to the original source.

flat response—A measure of a system's ability to reproduce audible frequencies while maintaining the same balance among them.

flutter—Very short rapid variations in the speed of the tape which cause changes in pitch and volume.

frequency—The repetition rate of sound or electrical energy. Bass frequencies range from 20 Hz to above 200 Hz; middle frequencies from 200 Hz to 3 kHz; high frequencies range from 3 kHz to 18 kHz.

frequency response—The measurement of the reproduction of frequencies over a specified range.

gain—The ratio of the output to the input of a system measured in dB.

gap—The opening between the poles of a head.

harmonic distortion—A form of distortion caused by the unwanted introduction of an overtone which is a multiple of another tone. Expressed in specifications as a percentage, the lower, the better.

head—The electromagnet across which the tape moves in a magnetic recorder/player.

impedance—The resistance to the flow of alternating current, given as "high" or "low" or in ohms so that a "match" can be made between two electronic components.

index counter—A counter which makes it easy to keep track of selections recorded on tape.

input—The provision for introducing a signal into an electronic component—usually a jack.

intermodulation distortion—A form of distortion formed by two or more tones. When given in specifications as a percentage, the lower the number, the better the equipment.

ips—Tape speed is measured in inches per second, which is abbreviated "ips."

jack—A receptacle for a plug.

kHz—Abbreviation for "kilohertz" or one thousand hertzs.

leader—A special nonmagnetic tape which can be spliced to either end of a magnetic tape to avoid damage.

level—The strength at which a signal is recorded or played back.

level indicator—A meter, neon bulb, or "magic eye" which shows the level or strength of a signal.

loop—A length of tape joined together at both ends to form a circle.

mil—One one-thousandth of an inch. Tape thickness is measured in mils.

mixer—An electronic device which combines two or more signals to be fed into a recorder.

monaural—A single channel recorder, as opposed to a stereophonic or multichannel recorder.

NAB Curve—The standard equalization playback curve set by the National Association of Broadcasters.

open-reel—Reel-to-reel tape carriers, as compared to cassettes or cartridges.

output—The signal delivered by an electronic device.

oxide—The ferro-magnetic particles coated on a base which form the magnetic part of the tape.

patch cord—A cord or cable connecting two pieces of electronic equipment.

pause control—A mode which permits stopping the player/recorder briefly for editing.

peak—The maximum instantaneous volume of any sound.

phasing—The establishment of correct polarity between output and speakers.

pinch roller—A roller or wheel which presses tape against the capstan shaft.

playback head—A magnetic head used to pick up sound recorded on tape.

pre-amplifier—An amplifier which changes the weak signal from a playback head to a level strong enough to drive the power amplifier.

pre-recorded tape—Tapes which are commercially available which are already recorded.

pressure pad—A felt pad which presses tape against the head.

print-through—The transfer of signals from layer to layer.

quadraphonic sound—Sound recorded and reproduced from four different directions.

rumble—A distortion caused by record players and existing in tape reproduction only if recorded originally.

saturation—The condition when a tape has reached its maximum magnetization.

separation—The degree to which two stereo signals are kept apart.

signal-to-noise-ratio—The difference between the loudest undistorted tone and the absence of a tone.

sound-on-sound—Recording a new signal with a previously recorded signal.

sound-with-sound—Listening to one channel while recording at the same time on another channel.

stereophonic sound—Two or more sound tracks played together to reproduce more realistic sound.

tape guides—Rollers which position the tape correctly.

tensilized polyester—A base which has been pre-stretched.

tone control—A device which permits changing tone from bass to treble.

track—The patch on which signals are placed in recording a magnetic tape.

unidirectional microphone—A microphone which picks up sounds coming from only one direction.

VU meter—A "volume unit" meter which indicates audio frequency power levels in dB.

wow—Variations in tape speed cause a form of distortion in pitch termed *wow*.

Useful Books on Tape Recorders

Anderson, Robert and Mark Mooney, *What You Should Know About Your Tape Recorder* (Boston: Herman Publishing, 1965).

Athey, Skipworth, *Magnetic Tape Recording* (Washington, D.C.: National Aeronautics and Space Administration, 1966).

Belt, Forest H., *One-Two-Three-Four Servicing Cassettes* (Indianapolis: Howard W. Sams and Co., Inc., 1972).

Belt, Forest H., *Pictorial Guide to Tape Recorder Repairs* (Blue Ridge Summit, Pa.: Tab Books, 1973).

Crowhurst, Norman H., *ABC's of Tape Recording* (Indianapolis: Howard W. Sams and Co., Inc., 1971).

Davidson, Homer, *Japanese Radio, Record and Tape Player Schematic Servicing Manual* (Blue Ridge Summit, Pa.: Tab Books, 1973).

How to Get the Most Out of Your Tape Recorder (Boston: Herman Publishing, 1965).

Jorgensen, Finn, *Handbook of Magnetic Recording* (Blue Ridge Summit, Pa.: Tab Books, 1970).

King, Gordon J., *Hi-Fi and Tape Recorder Handbook* (Levittown, New York: Transatlantic Arts, Inc., 1971).

Lloyd, J. M., *All in One Tape Recorder Book* (New York: Amphoto, 1970).

Lowman, Charles E., *Magnetic Recording* (New York: McGraw Hill, 1971).

Marshall, Robert, *Tape Recorder Service Manual and Trouble-Shooting Workbook* (New York: Amphoto, 1962).

Nijsen, C. G., *Tape Recorder* (New York: Drake Publishing, 1971).

Peters, Ken, *Modern Tape Recorders and Hi-Fi* (Levittown, New York: Transatlantic Arts, Inc., 1963).

Recording Basics (St. Paul, Minnesota: 3M, n.d.).

Rosenthal, Murray, *How to Select, Use, Troubleshoot Cassettes and Cartridge Recorders* (New York: Hayden Book Co., Inc., 1972).

Runstein, Robert E., *Modern Recording Techniques* (Blue Ridge Summit, Pa.: Tab Books, 1973).

Salm, Walter G., *Tape Recording for Fun and Profit* (Blue Ridge Summit, Pa.: Tab Books, 1969).

Sands, Leo G., *Questions and Answers About Auto Tape Units* (Indianapolis: Howard W. Sams and Co., Inc., 1973).

Schroder, H., *Tape Recording Servicing Mechanics* (New York: Van Nostrand Reinhold, 1968).

Tape Recording the Sounds of Your Life (Boston: Herman Publishing, 1965).

Westcott, Charles G. and Richard F. Dubbe, *Tape Recorders: How They Work*, Norman Crowhurst, ed. (Indianapolis: Howard W. Sams and Co., Inc., 1974).

Zuckerman, Arthur, *Tape Recording for the Hobbyist* (Indianapolis: Howard W. Sams and Co., Inc., 1973).

Directory of Manufacturers

Ampex Corporation
401 Broadway
Redwood City, California 94063

Astrocom/Marlux, Inc.
Box 370
Oneonta, New York 13820

Audio Devices, Inc.
100 Research Drive
Glenbrook, Connecticut 06906

Audio Magnetics Corporation
14600 S. Broadway
Gardena, California 90248

Audiotex
400 S. Wyman Street
Rockford, Illinois 61101

BASF Systems, Inc.
Crosby Drive
Bedford, Massachusetts 01730

BSR McDonald
Route 303
Blauvelt, New York 10913

Electro-Voice, Inc.
603 Cecil Street
Buchanan, Michigan 49107

Ercona Corporation
2121 Bellmore Ave.
Bellmore, New York 11701

General Electric Company
Syracuse, New Yrok 13201

Gladding-Claricon, Inc.
663 Dowd Ave.
Elizabeth, New Jersey 07201

Heath Company
Benton Harbor, Michigan 49022

Hervic Electronics. Inc.
1508 Cotner Ave.
Los Angeles, California 90025

JVC America, Inc.
50-35 56 Road
Maspeth, New York 11378

Koss Electronics, Inc.
4129 N. Point Washington Avenue
Milwaukee, Wisconsin 53212

Magnesonics Corporation
11036 Nestle Ave.
Northridge, California 91324

Maxell Corporation of America
130 W. Commercial Ave.
Moonachie, New Jersey 07074

Memorex
San Thomas at Central Expressway
Santa Clara, California 95052

Microtran Company, Inc.
145 E. Mineola Ave.
Valley Stream, New York 11582

Mura
50 S. Service Road
Jericho, New York 11753

Nagra Magnetic Recorders, Inc.
19 W. 44 Street
Room 715
New York, New York 10036

Nortronics Company, Inc.
8101 Tenth Avenue N.
Minneapolis, Minnesota 55427

Olson Electronics
260 S. Forge Street
Akron, Ohio 44308

Panasonic
Pan Am Building
200 Park Ave.
New York, New York 10017

Philco-Ford Corporation
Union Meeting Rd.
Blue Bell, Pennsylvania 19422

Pickering and Company, Inc.
101 Sunnyside Blvd.
Plainview, New York 11803

Radio Shack
2615 W. 7 Street
Fort Worth, Texas 76107

Royal Sound Company, Inc.
409 N. Main Street
Freeport, New York 11520

Russound/FMP, Inc.
Portsmouth Ave. Traffic Circle
Stratham, New Hampshire 03885

Sanyo Electric, Inc.
1200 West Walnut Street
Compton, California 90220

Sennheiser Electronics Corporation
10 West 37 Street
New York, New York 10018

Stanton Magnetics, Inc.
Terminal Drive
Plainview, New York 11803

Superex Electronics Corporation
151 Ludlow Street
Yonkers, New York 10705

Superscope, Inc.
8150 Vineland Ave.
Sun Valley, California 91352

Tandberg of America, Inc.
8 Third Ave.
Pelham, New York 10803

Tape-Athon Corporation
520 S. Isis Ave.
Inglewood, California 90301

TDK Electronics Corporation
755 Eastgate Blvd.
Garden City, New York 11530

Telex
9600 Aldrich Ave. S.
Minneapolis, Minnesota 55420

3M Company
3M Center
St. Paul, Minnesota 55101

Toshiba America, Inc.
40-06 Delong Street
Flushing, New York 11355

U.S. Pioneer Electronics Corporation
178 Commerce Road
Carlstadt, New Jersey 07072

Vidaire Electronics
Manufacturing Company
150 Buffalo Ave.
Freeport, New York 11520

Webcor Electronics
1815 Troy Street
New Albany, Indiana 47150

Yamaha International Corporation
P.O. Box 6600
Buena Park, California 90620

Zenith Radio Corporation
1900 N. Austin Ave.
Chicago, Illinois 60639

INDEX

A

B

C

D

E

F

G

H

I

J

L

M

N

O

P

Q

R

U

V

W

Z